MRTD (MULTI RESOLUTION TIME DOMAIN) METHOD IN ELECTROMAGNETICS

MRTD (Multi Resolution Time Domain) Method in Electromagnetics
Nathan A. Bushyager and Manos M. Tentzeris

ISBN: 978-3-031-00559-6 Bushyager/Tentzeris MRTD (Multi Resolution Time Domain) Method in Electromagnetics (paperback)
ISBN: 978-3-031-01687-5 Bushyager/Tentzeris MRTD (Multi Resolution Time Domain) Method in Electromagnetics (e-book)
Library of Congress Cataloging-in-Publication Data

First Edition
10 9 8 7 6 5 4 3 2 1

MRTD (MULTI RESOLUTION TIME DOMAIN) METHOD IN ELECTROMAGNETICS

Nathan A. Bushyager

Manos M. Tentzeris
NSF-PRC Associate Director for RF Research School of ECE
Georgia Institute of Technology,
Atlanta, GA 30332-250 U.S.A.

ABSTRACT

This book presents a method that allows the use of multiresolution principles in a time domain electromagnetic modeling technique that is applicable to general structures. The multiresolution time-domain (MRTD) technique, as it is often called, is presented for general basis functions. Additional techniques that are presented here allow the modeling of complex structures using a subcell representation that permits the modeling discrete electromagnetic effects at individual equivalent grid points. This is accomplished by transforming the application of the effects at individual points in the grid into the wavelet domain.

In this work, the MRTD technique is derived for a general wavelet basis using a relatively compact vector notation that both makes the technique easier to understand and illustrates the differences between MRTD basis functions. In addition, techniques such as the uniaxial perfectly matched layer (UPML) for arbitrary wavelet resolution and non-uniform gridding are presented. Using these techniques, any structure that can be simulated in Yee-FDTD can be modeled with in MRTD.

KEYWORDS

Adaptive Algorithms, MRTD, Wavelets, Adaptive Electromagnetic simulations, Time-/Space-adaptive gridding

Contents

CHAPTER 1

Introduction

The field of radio frequency (RF) and microwave design is growing at an astonishing rate due the confluence of several factors. Chief among these factors are the increased demand for RF and microwave consumer devices (cellular telephony and wireless data systems) and the increasing speed of digital devices, which has led to design limitations that previously applied only to traditional RF (radar and communications) circuits. There are also several quickly growing commercial RF applications such as automotive radar and RFID. As RF devices become more predominant in the consumer market, the design time for each generation decreases while performance demands increase. Design turnaround and performance gains traditional to the semiconductor device market are being expected of RF circuits. This in turn leads to increased demands on the RF designer and all associated tools, such as EM simulators.

Modern RF devices are built on a variety of technologies for a wide array of functionalities. In the attempt to reduce size and cost, multilayer substrates are commonplace. The constraints of each substrate are different, and the devices that can be fabricated with them often have no, or highly inaccurate, theoretical or empirical models. Design is usually performed using a top–down methodology, where the system is designed at the conceptual level and the details of the actual system are added as the design progresses. At the bottom of this design process, the actual physical layouts that represent the design blocks must be created. The characterization of these devices usually requires a full-wave electromagnetic simulator.

Full-wave simulators are often required to characterize effects that cannot be predicted, or properly accounted for, at higher levels of the design process. Examples of these effects are parasitic coupling, substrate modes, radiation, and package interference. These effects are often unique to the exact layout of a device and are too complicated to be treated theoretically. Any simulator that models the complete physics of electromagnetic interaction can be used to model these devices; however, time-domain techniques are popular and particularly well suited to these devices.

Time-domain techniques, as the name implies, determine the electromagnetic fields in a structure in response to a source condition. They are contrasted to frequency-domain methods, which determine the response to a harmonic source. There are several advantages to each class of methods, and the requirements of a particular problem determine which is best. The characteristic that is most often cited in recommending time-domain methods for the simulation of microwave devices is that a broadband response can be determined from a single simulation. In addition, time-domain simulators are usually derived using a differential form of Maxwell's equations, relating fields at points and allowing easy discretization of complex structures; they do not require the calculation of Green's functions or the inversion of a matrix. Furthermore, nonlinear effects can be simulated easily, as the field strength is measured as a function of time and these effects can be directly applied. The two main drawbacks of time-domain techniques are representing dispersive media and speed. The speed of these techniques, while comparable to other full-wave techniques, is not nearly fast enough for modern microwave design.

The most mature and widely used time-domain simulation technique is finite-difference time-domain (FDTD) (1). The Yee-FDTD scheme is one of the oldest, and one of the simplest, time-domain EM simulation techniques. It has remained popular due to this simplicity; not only is it relatively easy to apply, but it is also highly flexible. Because of its flexibility, the method has been applied to practically every type of electromagnetic analysis, from radar-cross-section investigation to optical characterization. A number of compatible techniques have been developed to model specific effects in the method. When developing other methods, the Yee-FDTD scheme is often used as a benchmark; most other techniques are designed to be faster.

While a number of techniques have been developed that are more efficient than FDTD for the same accuracy, these techniques add complexity and cannot be as generally applied. Several FDTD techniques using higher order basis functions (than the pulse functions used in FDTD) have been presented, but it is difficult to model general microwave structures because of the difficulty of representing boundary conditions (such as those required to model arbitrarily positioned metal structures). While the design community desires an improved method that offers several orders of magnitude improvement in simulation time, this is not possible. One promising method of decreasing design time is to identify those parts of the circuit that require full-wave simulation, and characterizing the remainder of the circuit with less computationally intensive techniques. However, as integration increases, the areas of the circuit that require full-wave simulation will increase. Thus, it is necessary to use the most efficient full-wave technique possible. One technique that has been suggested is multiresolution time-domain (MRTD) (2).

MRTD uses a wavelet-based discretization to represent the EM fields. The wavelets allow the resolution of the simulation to be changed as a function of both time and space. If the resolution is changed to respond to a propagating waveform, the number of operations that must be performed to characterize a structure can be minimized. The MRTD technique has been referred to as a generalization of the FDTD technique. The derivation of the method follows the same general scheme, but the relatively simple pulse basis functions used in the FDTD technique are replaced with the wavelet functions.

The purpose of this monograph is to present the MRTD technique to a reader unfamiliar with multiresolution (wavelet) analysis and little or no experience in numerical modeling. A more advanced reader will also benefit from the treatment presented here, especially the later chapters where newer techniques in MRTD analysis and examples are presented. In addition, the notation used to present

MRTD in this book attempts to unify the methods used to present all wavelet bases. By use of this method, a compact representation is presented that allows the characteristics of various wavelet schemes to be compared. This makes the method both easier to understand and provides a straightforward method to deal with the large number of nested coefficients that appear in a multiresolution scheme.

Since this is a relatively short book, it cannot cover all aspects of the complex multiresolution time domain technique. A multitude of papers have been written discussing the technique, using a wide variety of wavelet basis functions. The aspects of each wavelet scheme are well beyond the scope of this presentation. Because of this limitation, the lecture is organized such that the reader is presented with the general MRTD technique in the first chapter and overview of its properties and a discussion of its implementation in the second chapter. The remaining chapters focus on the implementation of the MRTD technique using Haar basis functions. The reason for this is twofold. First, the Haar wavelets are the simplest that can be used and thus can be covered in detail in the relatively short length of this book. Second, the Haar wavelets, like Yee-FDTD, are extremely general, and can thus be easily applied to complex structures while still keeping the advantages of the MRTD technique. The main advantage of the Haar wavelets is the ability to represent local features, such as perfect conductors, without the need to use image theory. A presentation using the Haar wavelets gets the reader familiar with the MRTD technique quickly, while still presenting the framework for a powerful, adaptive simulator.

C H A P T E R   2

Background

This monograph presents techniques that can be used to model complex microwave structures in MRTD. In this section, the derivation of the MRTD technique in three dimensions for a general wavelet basis is presented. In addition, as the MRTD technique is presented as an adaptive alternative to FDTD, and several MRTD examples will be contrasted to FDTD, a brief overview of the FDTD method is presented. This chapter begins with a discussion of MRTD basics, including wavelet basis functions and the method of moments, followed by a discussion of two specific areas, PML and PEC modeling. After these topics are addressed, an overview of FDTD, focusing on its similarities and differences with MRTD, is presented.

2.1   MRTD BASICS

2.1.1   Wavelet Overview

The MRTD technique was first presented by Krumpholz and Katehi in 1996 [2]. At that time multiresolution analysis, the application of wavelet bases to numerical problems, was becoming popular in a number of fields as a way of increasing both the efficiency and accuracy of numerical methods. Their original MRTD paper provides a general discretization technique that can be used with any wavelet basis; however, the paper focused on the Battle–Lemarie wavelet scheme. Several other wavelet schemes have since been applied in the same manner, including Haar [3], Daubechies [4], and Cohen–Daubechies–Feauveau (CDF) [5].

While the above-mentioned wavelet schemes are all very different, they share a number of characteristics that make them suitable for numerical modeling. Most wavelet schemes used in for MRTD are orthonormal and are characterized by a scaling function and a mother wavelet. The mother wavelet is generated using the scaling function and is in turn used to generate all other wavelets that constitute the basis. The orthogonality that exists between the wavelets and scaling function, as well as all wavelets with other wavelets, make them natural choices for numerical discretization.

The key concept regarding wavelet expansions is that of *levels* of wavelet resolution. These levels correspond to sets of functions that can be added to the expansion to increase the accuracy of the wavelet discretization. In the following expressions, the scaling functions will be represented as $\varphi_i(x)$ [6], where

$$\varphi_i(x) = \varphi\left(\frac{x}{\Delta x} - i\right) \tag{1}$$

and $\varphi(x)$ represents the general scaling function. Likewise, the wavelets are offset throughout the grid, but are characterized with three indices. A typical wavelet coefficient, as used in MRTD, is represented as

$$\psi_{i,p}^r(x) = 2^{r/2}\psi_0\left(2^{r/2}\left(\frac{x}{\Delta x} - i\right) - p\right). \tag{2}$$

In this function, r represents the wavelet resolution and p can take any integer value between 0 and $2^r - 1$. Using these coefficients, each level of resolution, r, contains 2^r wavelets, offset by $\Delta x / 2^r$.

For all wavelet schemes used in multiresolution analysis, the following hold [6]:

$$\int \varphi_i(x)\varphi_j(x) = \delta_{i,j} \tag{3}$$

$$\int \varphi_i(x)\psi_{j,p}^r(x) = 0 \quad \forall i, j, r, p \tag{4}$$

$$\int \psi_{i,p}^r(x)\psi_{j,q}^s(x) = \delta_{i,j}\delta_{r,s}\delta_{p,q}, \tag{5}$$

where $\delta_{i,j}$ is the Kronecker delta function

$$\delta_{i,j} = \begin{cases} 1 & i = j \\ 0 & i \neq j \end{cases}. \tag{6}$$

If the closed subspace represented by all wavelets of resolution r and higher is termed V_r, the following properties are required:

$$\ldots V_{-3} \subset V_{-2} \subset V_{-1} \subset V_0 \subset V_1 \subset V_2 \subset V_3 \ldots \tag{7}$$

$$\overline{\bigcup_{i \in Z} V_i} = L^2(R) \tag{8}$$

$$\overline{\bigcap_{i \in Z} V_i} = 0, \tag{9}$$

where Z is the set of all integers and $L^2(R)$ is the set of all square integrable functions. This is by no means a complete description of wavelets; it is only what is necessary to discuss the concepts in this monograph. Several books, such as [6], have been written on wavelet analysis, these can provide a much more broad appreciation of multiresolution analysis.

The consequence of these properties is that each addition of a level of wavelet resolution increases the accuracy of the representation, and arbitrary accuracy can be achieved using the appropriate wavelet resolution. This is in contrast to methods of improving accuracy in other basis functions, where increased accuracy is achieved by contracting the domain and increasing the number of basis functions. There are several basis functions that can be used to meet these requirements.

2.1.2 Haar Wavelets

The oldest and simplest wavelet basis is the Haar basis functions. When these wavelets are used in MRTD, a scheme equivalent to FDTD can result [7]. However, when used correctly, a time and space adaptive scheme that is more numerically efficient than FDTD results. As mentioned above, a number of other wavelet schemes have been used with the MRTD technique.

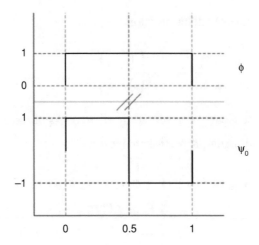

FIGURE 1: Haar scaling function (top) and mother wavelet (bottom).

An overview of the Haar wavelets is presented here. They are simple enough that they can be used to explain wavelet concepts relatively easily in one and two dimensions, while adequately demonstrating wavelet properties (three-dimensional (3-D) wavelets will be discussed, but are difficult to present in a two-dimensional (2-D) format). The examples presented in this work use Haar wavelets exclusively; they are preferred for a number of reasons that will be discussed. In addition, the composite cell technique presented in later chapters is practical only for Haar wavelets.

The Haar scaling function φ and mother wavelet ψ are presented in Fig. 1. The Haar scaling function is defined as $\chi_{[0,1)}$, the characteristic function from zero to one. The mother wavelet is based on this function. It is defined as

$$\psi_0(x) = \varphi(2x) - \varphi(2(x - \tfrac{1}{2})), \tag{10}$$

Higher level wavelets can be generated using (2). These wavelets for resolutions 1 and 2 are presented in Fig. 2.

As stated previously, and demonstrated in Fig. 2, there are 2^r wavelets for each level of resolution. Thus, there are 2 wavelets at $r = 1$ and 4 wavelets at $r = 2$. These functions represent scaled and translated mother wavelets. The scaling factor,

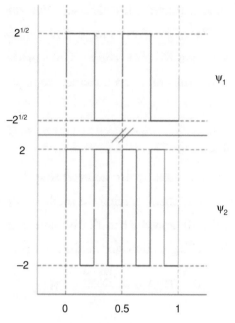

FIGURE 2: Haar wavelets for resolution 1 and 2.

$2^{r/2}$, presented in (2) is required to orthonormalize the functions, which guarantees (5). The domain of all of the wavelet functions for any level of resolution is identical to that of the mother wavelet and scaling function. A scheme that uses wavelets up to and including a resolution level r will use 2^{r+1} functions (including the scaling function).

There are several advantages to using Haar wavelets to model electromagnetic phenomena. The first is their finite domain nature. When Haar wavelets are used to expand an electric or magnetic field, the scaling functions and wavelets from one cell to the next do not overlap. This allows hard boundary conditions (setting discrete areas to a fixed value) to be easily applied. For example, a perfect-electrical-conductor (PEC) boundary condition can be applied by zeroing tangential electric fields. If the basis functions were to overlap, this condition could not be applied. The second advantage of using Haar wavelets is the relative ease of performing derivatives and integrals (although these integrals can be tabulated and thus calculated only once for any number of simulations). Because of their pulse nature, these operations become simple arithmetic. It will be seen that this eases the generation of the MRTD scheme. The major disadvantage of Haar wavelets is in the numerical dispersion of the resultant MRTD scheme. Compared to most other wavelet schemes, denser grids, and thus more coefficients, are required for Haar-MRTD schemes.

2.1.3 MRTD Scheme

The MRTD scheme is generated by first representing the electric and magnetic fields in terms of wavelets and then applying the method of moments to Maxwell's curl equations, (11) and (12), and the constitutive equations, (15) and (16). The differential form of Maxwell's equations in the time domain can be expressed as

$$\nabla \times \mathbf{E}\left(t\right) = -\frac{\partial \mathbf{B}\left(t\right)}{\partial t} - \mathbf{M}\left(t\right) \tag{11}$$

$$\nabla \times \mathbf{H}\left(t\right) = \frac{\partial \mathbf{D}\left(t\right)}{\partial t} + \mathbf{J}\left(t\right) \tag{12}$$

$$\nabla \cdot \mathbf{D}\left(t\right) = \rho_{\mathrm{e}}\left(t\right) \tag{13}$$

$$\nabla \cdot \mathbf{B}\left(t\right) = \rho_{\mathrm{m}}\left(t\right), \tag{14}$$

where $\mathbf{E}$ is the electric field (V/m), $\mathbf{H}$ is the magnetic field (A/m), $\mathbf{D}$ is the electric flux density (C/m^2), $\mathbf{B}$ is the magnetic flux density (Wb/m^2), ρ is charge density (C/m^3), $\mathbf{M}$ is the magnetic current density (V/m^2), and $\mathbf{J}$ is the electric current density (A/m^2). These equations relate the electric and magnetic fields at points in space. This analysis is limited to isotropic, nondispersive media, which have the constitutive equations

$$\mathbf{D}\left(\mathbf{r}\right) = \varepsilon\left(\mathbf{r}\right)\mathbf{E}\left(\mathbf{r}\right) \tag{15}$$

$$\mathbf{B}\left(\mathbf{r}\right) = \mu\left(\mathbf{r}\right)\mathbf{H}\left(\mathbf{r}\right), \tag{16}$$

where $\mathbf{r}$ is the position vector.

For a general material interface, the boundary conditions are

$$\hat{\mathbf{n}} \cdot \left(\mathbf{D}_2 - \mathbf{D}_1\right) = \rho_{\mathrm{s}} \tag{17}$$

$$\hat{\mathbf{n}} \cdot \left(\mathbf{B}_2 - \mathbf{B}_1\right) = 0 \tag{18}$$

$$\hat{\mathbf{n}} \times (\mathbf{E}_2 - \mathbf{E}_1) = 0 \tag{19}$$

$$\hat{\mathbf{n}} \times (\mathbf{H}_2 - \mathbf{H}_1) = \mathbf{J}_s, \tag{20}$$

where ρ_s, and $\mathbf{J}_s$ are the charge and electric current densities on the interface and $\hat{\mathbf{n}}$ is the normal unit vector to the interface.

Faraday's and Ampere's laws, (11) and (12) respectively, are vector equations and are expressed in Cartesian coordinates as

$$\frac{\partial B_x}{\partial t} = \frac{\partial E_y}{\partial z} - \frac{\partial E_z}{\partial y} - M_x \tag{21}$$

$$\frac{\partial B_y}{\partial t} = \frac{\partial E_z}{\partial x} - \frac{\partial E_x}{\partial z} - M_y \tag{22}$$

$$\frac{\partial B_z}{\partial t} = \frac{\partial E_x}{\partial y} - \frac{\partial E_y}{\partial x} - M_z. \tag{23}$$

$$\frac{\partial D_x}{\partial t} = \frac{\partial H_z}{\partial y} - \frac{\partial H_y}{\partial z} - J_x \tag{24}$$

$$\frac{\partial D_y}{\partial t} = \frac{\partial H_x}{\partial z} - \frac{\partial H_z}{\partial x} - J_y \tag{25}$$

$$\frac{\partial D_z}{\partial t} = \frac{\partial H_y}{\partial x} - \frac{\partial H_x}{\partial y} - J_z. \tag{26}$$

To generate the MRTD scheme, each E and H field in Faraday's and Ampere's laws must be expanded in terms of scaling and wavelet functions. A one-dimensional (1-D) field, for example E_x in a 1-D scheme, is expressed in scaling and wavelet functions up to resolution r_{max} as

$$E_x(x) = \sum_{n,i} h_n(t) \left[{}_n E_i^{x,\phi} \varphi_i(x) + \sum_{r=0}^{r_{max}} \sum_{p=0}^{2^r-1} {}_n E_{i,r,p}^{x,\psi} \psi_{i,p}^r(x) \right], \tag{27}$$

where ${}_n E_i^{x,\phi}$ and ${}_n E_{i,r,p}^{x,\psi}$ are coefficients representing the magnitudes of the scaling and wavelet functions. This function is the discretization in both time and space. The time discretization, $h_n(t)$, is performed with simple pulse functions, equivalent to the Haar scaling functions presented in Fig. 1. The pulse functions are used in time to ensure causality. If the functions representing each time step overlapped, it would be possible for past events to be modified by future ones. The spatial index, i, indicates the position in cells along the x axis. The cells are defined by the domain of each scaling function, Δx. It is, technically, possible to use one scaling function the size of the entire domain. Practically, however, this is difficult to implement, and it removes some of the ability to locally increase resolution. In practice, several cells per the maximum excited wavelength are used. The criteria for choosing the cell size are discussed later, as a part of numerical dispersion.

In order to expand the fields in multiple dimensions, the products of all scaling/wavelet functions of each direction must be used. In two dimensions, this means four groups of coefficients, scaling-x/scaling-y, wavelet-x/scaling-y, scaling-x/wavelet-y, and wavelet-x/wavelet-y coefficients. As in (27), the wavelet terms require sums, and the wavelet/wavelet terms have nested x and y sums. For example, in a 2-D simulation, with $r_{max} = 0$, the four functions used to represent the fields for the Haar wavelets are depicted in Fig. 3.

The maximum resolution is not required to be the same in each direction. For any wavelet basis, the number of coefficients required for a given maximum resolution is

$$\text{number of coefficients} = 2^{D + \sum\limits_{i=x,y,z} r_{max,i}}, \tag{28}$$

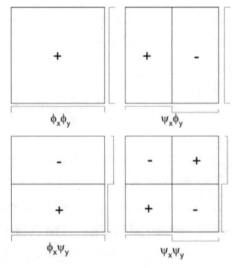

FIGURE 3: Two-dimensional Haar coefficients for $r_{max} = 0$.

where D is the dimensionality of the simulator and the sum indicates that the maximum resolution may vary by direction. In a 3-D simulator, there are nine groups of coefficients needed to represent the field expansion, and an increase in resolution in every direction increases the number of coefficients needed by three powers of two.

The expansion of the E_x field in three dimensions, for any wavelet basis, is

$$E_x(x) = \sum_n b_{n(t)} \sum_{i,j,k} \left[\begin{array}{l} {}_nE_{i,j,k}^{x,\phi\phi\phi}\,\varphi_i(x)\varphi_j(y)\varphi_k(z)+ \\[6pt] \sum_{r=0}^{r_{max}}\sum_{p=0}^{2^r-1} {}_nE_{i,j,k,r,p}^{x,\psi\phi\phi}\,\psi_{i,p}^r(x)\varphi_j(y)\varphi_k(z)+ \\[6pt] \sum_{r=0}^{r_{max}}\sum_{p=0}^{2^r-1} {}_nE_{i,j,k,r,p}^{x,\phi\psi\phi}\,\varphi_i(x)\psi_{j,p}^r(y)\varphi_k(z)+ \\[6pt] \sum_{r=0}^{r_{max}}\sum_{p=0}^{2^r-1} {}_nE_{i,j,k,r,p}^{x,\phi\phi\psi}\,\varphi_i(x)\varphi_k(y)\psi_{k,p}^r(z)+ \\[6pt] \sum_{r=0}^{r_{max}}\sum_{p=0}^{2^r-1}\sum_{s=0}^{r_{max}}\sum_{q=0}^{2^s-1} {}_nE_{i,j,k,r,p,s,q}^{x,\psi\psi\phi}\,\psi_{i,p}^r(x)\psi_{j,q}^s(y)\varphi_k(z)+ \\[6pt] \sum_{r=0}^{r_{max}}\sum_{p=0}^{2^r-1}\sum_{s=0}^{r_{max}}\sum_{q=0}^{2^s-1} {}_nE_{i,j,k,r,p,s,q}^{x,\psi\phi\psi}\,\psi_{i,p}^r(x)\varphi_j(y)\psi_{k,q}^s(z)+ \\[6pt] \sum_{r=0}^{r_{max}}\sum_{p=0}^{2^r-1}\sum_{s=0}^{r_{max}}\sum_{q=0}^{2^s-1} {}_nE_{i,j,k,r,p,s,q}^{x,\phi\psi\psi}\,\varphi_i(x)\psi_{j,p}^r(y)\psi_{k,q}^s(z)+ \\[6pt] \sum_{r=0}^{r_{max}}\sum_{p=0}^{2^r-1}\sum_{s=0}^{r_{max}}\sum_{q=0}^{2^s-1}\sum_{v=0}^{r_{max}}\sum_{w=0}^{2^v-1} {}_nE_{i,j,k,r,p,s,q,v,w}^{x,\psi\psi\psi}\,\psi_{i,p}^r(x)\psi_{j,q}^s(y)\psi_{k,w}^v(z) \end{array} \right].$$

$$(29)$$

The eight groups of coefficients represent all of the combinations of scaling/wavelet functions in each direction. Using (28), for $r_{max} = 0$ in all directions, there are eight coefficients per cell. For $r_{max} = 2$, there are 512 coefficients per cell. The field coefficients are identified in this equation by the superscripts, denoting the field direction and scaling/wavelet component for each direction. For example, $E^{y,\phi\phi\psi}$ denotes a y-directed electric field coefficient, with a scaling function used in the x and y directions, and a wavelet in the z direction. The coefficients have a maximum of nine indices (for the $E^{dir,\psi\psi\psi}$ coefficient), which represent the position in space of the scaling function as well as the resolution and position of each wavelet.

The 3-D MRTD scheme is generated by first representing all fields (E, H, D, and B) in scaling/wavelet function expansions, as in (29), and setting the time and space steps. However, the fields are not collocated in space or time. The original MRTD scheme presented in [2] offsets electric fields one half a cell (the offset between scaling functions) along their coordinate axis (E_x by $\Delta x/2$ in x, for example), while magnetic fields are offset by half a cell in their two normal directions (H_x by $\Delta y/2$ in y and $\Delta z/2$ in z). However, a correction to this field arrangement [8,9] was later made so that that the magnitude of the offset depends on the maximum wavelet resolution used in the simulation. The corrected offset, s_d, is

$$s_d = \frac{\Delta d}{2^{r_{d,\max}+2}},\tag{30}$$

where d denotes the direction (x, y, or z). The purpose of this offset is discussed in more detail as a consequence of numerical dispersion, but a brief graphical discussion is now presented which explains some of the purpose of the choice of offset.

Fig. 4 shows a 2-D Haar cell with $r_{\max} = 0$. The $r = 0$ wavelets are shown as a reference. The MRTD technique is characterized by cells that contain several basis functions. As a consequence, the field variation throughout the cell is not limited to the shape of a single function, but rather the sum of several. The position of the wavelet coefficients at $r_{\max}$, which can be calculated using (2), defines a number of *equivalent grid points* [9]. Specifically, two equivalent grid points are

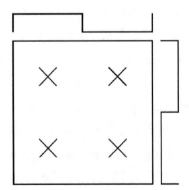

FIGURE 4: Two-dimensional Haar cell, $r_{\max} = 0$, equivalent grid points.

defined per r_{max} level wavelet positioned $\Delta x/(4 \times 2^{r_{max}})$ and $3\Delta x/(4 \times 2^{r_{max}})$ from the position specified in (2). It should be noted that the number of equivalent grid points is equal to the number of basis functions, (28). While the half cell offset used in [2] was chosen to give the cells an offset equal to Yee-FDTD [1], the offset presented in (30) arranges the equivalent grid points in the Yee manner. It should be noted that while Fig. 4 is shows the wavelets for the Haar scheme, the position of the equivalent grid points is valid for any wavelet basis. Furthermore, the cell size represents the domain of the scaling function in the Haar case, but it represents the spacing between the centers of the scaling functions in the general case. The equivalent grid points in Fig. 4 are marked with ×.

If a 2-D scheme with E_x, E_y, and H_z (TE$_z$ mode) is simulated with the $r_{max} = 0$ cell of Fig. 4, and the fields are offset using (30) in the manner suggested above, the field arrangement shown in Fig. 5 results. This figure shows the equivalent grid points for all three fields and is valid for any wavelet basis. The arrangement of the grid points in this case is the same as the Yee-FDTD cell. If the same

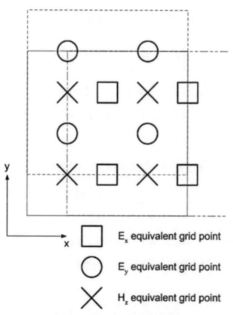

FIGURE 5: Equivalent grid points for a 2-D MRTD cell, $r_{max} = 0$.

arrangement is used for a full 3-D scheme, the equivalent grid points are arranged in the same manner as Yee-FDTD.

Like FDTD, the update equations in MRTD are fully explicit. Thus, the time step must be chosen according to a stability requirement. This is discussed later. Also like FDTD, the electric and magnetic fields are offset in time. The time basis functions used for most MRTD schemes are pulses, as in FDTD. Similarly, the fields are offset by half a time step. Once the field offset is determined, the MRTD update scheme can be determined using the method of moments.

2.1.4   Method of Moments Overview

The method of moments is a general method used in a variety of numerical schemes. A brief overview of techniques that relate to MRTD is presented here. First, the method is applied by representing the unknown function as a sum of unknown coefficients multiplied by known basis functions. Next, a series of testing functions are chosen. These functions are multiplied into both sides of the equation and the integral is taken over the entire domain. This process is referred to as an inner product and the following notation will be used:

$$\langle f, g \rangle = \int f(\mathbf{r}) g(\mathbf{r}) \partial \mathbf{r}. \tag{31}$$

If the number of testing functions is equal to the number of unknown coefficients, the result is a series of linear equations that can be solved to determine the coefficients.

As a simple 1-D (in space) example, the equation

$$f(x) = \frac{\partial b(x)}{\partial x} \tag{32}$$

is presented, where $f(x)$ is unknown and $b(x)$ is a known function. The unknown expression can be expanded into a sum of basis functions

$$f(x, t) = \sum_{n=0}^{N} a_n c_n(x), \tag{33}$$

where a_n are unknown scalar coefficients and $c_n(x)$ are the known basis functions. A set of testing function $w_n(x)$ is chosen, and the inner product of (32) with each

$$\langle w_n, f \rangle = \langle w_n, b' \rangle \tag{34}$$

is taken. The result is a system of equations,

$$\begin{bmatrix} a_1 \langle w_1, c_1 \rangle & a_2 \langle w_1, c_2 \rangle & \cdots & a_N \langle w_1, c_N \rangle \\ a_1 \langle w_2, c_1 \rangle & a_2 \langle w_2, c_2 \rangle & & \\ \vdots & & \ddots & \\ a_1 \langle w_N, c_1 \rangle & & & a_N \langle w_N, c_N \rangle \end{bmatrix} = \begin{bmatrix} \langle w_1, b' \rangle \\ \langle w_2, b' \rangle \\ \vdots \\ \langle w_N, b' \rangle \end{bmatrix}, \tag{35}$$

which can be solved to yield the unknown coefficients, a_n. If the testing functions are chosen such that

$$\langle w_m, c_n \rangle = \delta_{m,n}, \tag{36}$$

then the matrix on the left-hand side of (35) is diagonal. In this case, the scheme is explicit; a matrix does not need to be inverted to solve the equations. Each coefficient can be determined directly,

$$a_n = \langle w_n, b' \rangle. \tag{37}$$

If an orthonormal set of basis functions is chosen

$$\langle w_m, w_n \rangle = \delta_{m,n}, \tag{38}$$

then the basis functions can be used as testing functions. This technique is called Galerkin's method.

2.2 MRTD UPDATE SCHEME

2.2.1 Time Localization

MRTD update equations are determined using Galerkin's method with wavelet discretizations of the electric and magnetic fields, such as (29). The half time step offset of the electric and magnetic field ensure that the fields are always updated from values at previous time steps, and thus the only unknowns are the updated values of the fields. Using (21)–(23), B fields are determined from previous E field values. Likewise, (24)–(26) are used to determine D fields from previous H fields. In many FDTD schemes the constitutive relationship, (15) and (16), is applied discretely at each cell, and thus H values are determined directly from E values, and vice-versa. For general MRTD basis functions, which can extend over the entire domain, this is not possible. Furthermore, even for non-overlapping, finite-domain wavelets, such as Haar wavelets, it is not practical because the dielectric constant cannot vary over the relatively large MRTD cell. Thus, the method of moments must also be applied to (15) and (16), to determine E fields from D fields and H fields from B fields. While this does add complexity to the method, it also allows the material to vary continuously throughout the cell.

The MRTD update equations are determined by localizing the coefficients in time by testing with the time basis functions (pulses) and then localizing in space by testing with the scaling/wavelet functions. The time derivative of the pulse function yields two Dirac Delta functions, located at the edges of the pulse,

$$\frac{\partial h_n(t)}{\partial t} = \delta(t - n\Delta t) - \delta(t - (n + 1)\Delta t). \tag{39}$$

The time derivatives of the pulse functions that make up the time discretization are a Delta train, represented by the top line of Fig. 6. In (21)–(26), one field value is differentiated in time, while the other field values are differentiated in space. As stated previously, the E and H fields are offset in time by half a time step, which means that the spatial derivative terms are offset in time by half a time step from the time derivative terms.

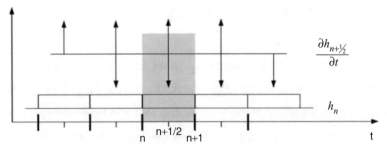

FIGURE 6: Time localization using pulse functions.

The time-testing function is chosen to overlap with the terms in the spatial derivative. If the B (and H) coefficients are located at the $n + 1/2$ time steps and the D (and E) functions are located at the n time steps, Fig. 6 represents the time discretization of (21)–(23). The delta train on the top of the figure represents the time derivative of the B fields, while the pulse functions on the bottom represent the E fields. The shaded area represents the domain of a testing function, which is collocated in time with the E fields. If the time and space basis functions that represent the E fields are separated

$$E(\mathbf{r}, t) = \sum_{n=0}^{N} h_n(t)\, {}_n E(\mathbf{r}), \tag{40}$$

where ${}_n E(\mathbf{r})$ is the wavelet/scaling space discretization for time step n, the quantities in the brackets of (29). Similarly,

$$B(\mathbf{r}, t) = \sum_{n=0}^{N} h_{n+1/2}(t)\, {}_{n+1/2} B(\mathbf{r}), \tag{41}$$

represents the B fields. Using these expressions, the inner products with h_n are

$$\langle h_n(t),\, E(\mathbf{r}, t) \rangle = \Delta t\, {}_n E(\mathbf{r}), \tag{42}$$

$$\left\langle h_n(t),\, \frac{\partial B(\mathbf{r}, t)}{\partial t} \right\rangle = {}_{n+1/2} B(\mathbf{r}) - {}_{n-1/2} B(\mathbf{r}). \tag{43}$$

As an example, (42) and (43) can be applied to (21). Ignoring the current terms and taking the inner product with h_n

$$\left\langle h_n,\, \frac{\partial B_x}{\partial t} \right\rangle = \left\langle h_n,\, \frac{\partial E_y}{\partial z} - \frac{\partial E_z}{\partial y} \right\rangle \tag{44}$$

yields

$$_{n+1/2} B_x(\mathbf{r}) - {}_{n-1/2} B_x(\mathbf{r}) = \Delta t \left[\frac{\partial\, {}_n E_y(\mathbf{r})}{\partial z} - \frac{\partial\, {}_n E_z(\mathbf{r})}{\partial y} \right]. \tag{45}$$

Similar expressions can be found using (22) to (26). Solving for $_{n+1/2} B_x(\mathbf{r})$,

$$_{n+1/2} B_x(\mathbf{r}) = {}_{n-1/2} B_x(\mathbf{r}) + \Delta t \left[\frac{\partial\, {}_n E_y(\mathbf{r})}{\partial z} - \frac{\partial\, {}_n E_z(\mathbf{r})}{\partial y} \right] \tag{46}$$

yields an update equation for $_{n+\frac{1}{2}}B_x(\mathbf{r})$ in terms of quantities from the previous time step. The next step in determining the MRTD update equations is to test with the spatial scaling/wavelet functions.

2.2.2 Space Localization

The 3-D basis functions are expressed as $\Gamma(x)\,\Gamma(y)\,\Gamma(z)$, where each Γ can be either φ_l or $\psi_{l,p}^r$, with l representing a directional index, i, j, or k. As stated in (28), for maximum resolution in the x, y, z direction $r_{\max,x}$, $r_{\max,y}$, $r_{\max,z}$, respectively, each cell (denoted by the triplet i, j, k) contains $2^{3+r_{\max,x}+r_{\max,y}+r_{\max,z}}$ basis functions. To determine an update formula for each coefficient, the testing functions are collocated in space with the time differentiated functions in Faraday's and Ampere's laws. In (46) this is equivalent to collocating the scaling/wavelet testing functions with the B_x fields.

A convenient notation was first introduced in [8], which represents the field discretization as a vector. Here, this notation is generalized for any wavelet basis, which allows the MRTD update equations to be written in a compact and easily understandable form. The wavelet/scaling coefficients for all wavelet resolutions can be represented as a vector, for example

$$_{n+\frac{1}{2}}\mathbf{B}_{x,i,j,k} = \begin{bmatrix} _{n+\frac{1}{2}}B_{i,j,k}^{x,\phi\phi\phi} \\ _{n+\frac{1}{2}}B_{i,j,k,0,0}^{x,\psi\phi\phi} \\ \vdots \\ _{n+\frac{1}{2}}B_{i,j,k,r_{\max,z},2^{r_{\max,z}}-1}^{x,\phi\phi\psi} \\ _{n+\frac{1}{2}}B_{i,j,k,0,0,0,0}^{x,\psi\psi\phi} \\ \vdots \\ _{n+\frac{1}{2}}B_{i,j,k,r_{\max,y},2^{r_{\max,y}}-1,r_{\max,z},2^{r_{\max,z}}-1}^{x,\phi\psi\psi} \\ _{n+\frac{1}{2}}B_{i,j,k,0,0,0,0,0,0}^{x,\psi\psi\psi} \\ \vdots \\ _{n+\frac{1}{2}}B_{i,j,k,r_{\max,x},2^{r_{\max,x}}-1,r_{\max,y},2^{r_{\max,y}}-1,r_{\max,z},2^{r_{\max,z}}-1}^{x,\psi\psi\psi} \end{bmatrix}. \tag{47}$$

If another vector Γ is defined

$$
\Gamma_{i,j,k} = \begin{bmatrix}
\varphi_i(x)\varphi_j(y)\varphi_k(z) \\
\psi^0_{i,0}(x)\varphi_j(y)\varphi_k(z) \\
\vdots \\
\varphi_i(x)\varphi_k(y)\psi^{r_{max,z}}_{k,2^{r_{max,z}}-1}(z) \\
\psi^0_{i,0}(x)\psi^0_{j,0}(y)\varphi_k(z) \\
\vdots \\
\varphi_i(x)\psi^{r_{max,y}}_{j,2^{r_{max,y}}-1}(y)\psi^{r_{max,z}}_{k,2^{r_{max,z}}-1}(z) \\
\psi^0_{i,0}(x)\psi^0_{j,0}(y)\psi^0_{k,0}(z) \\
\vdots \\
\psi^{r_{max,x}}_{i,2^{r_{max,x}}-1}(x)\psi^{r_{max,y}}_{j,2^{r_{max,y}}-1}(y)\psi^{r_{max,z}}_{k,2^{r_{max,z}}-1}(z)
\end{bmatrix},
\tag{48}
$$

then

$$
_{n+1/2}B_x(\mathbf{r}) = \sum_{i,j,k} \Gamma^T_{i,j,k}{}_{n+1/2}\mathbf{B}_{x,i,j,k}.
\tag{49}
$$

The update equation for each coefficient of $\mathbf{B}$ can then be determined by taking the inner product of (46) with each scaling/wavelet coefficient (every row of (48)).

Because the basis functions are orthonormal, a separate update equation is found for each scaling/wavelet coefficient. However, because of the offset of the B and E fields, the scaling/wavelet functions used for the B fields are not orthogonal with the functions used for the E fields. If the basis functions representing the B_x field are offset in the positive y and z directions from cell i, j, k (located at $i\Delta x$, $j\Delta y$, $k\Delta z$), while the E_y fields are offset in the y direction and E_z is offset in the z direction, then, using the vector notation of (47), the update for (46) becomes

$$
_{n+1/2}\mathbf{B}_{x,i,j,k} = {}_{n-1/2}\mathbf{B}_{x,i,j,k} + \frac{\Delta t}{\Delta x \Delta y \Delta z}\left[\sum_m \mathbf{U}^{B_x}_{E_y,m}{}_n\mathbf{E}_{y,i,j,k+m}\right.
$$
$$
\left. + \sum_m \mathbf{U}^{B_x}_{E_z,m}{}_n\mathbf{E}_{z,i,j+m,k}\right].
\tag{50}
$$

In this expression, $\mathbf{U}$ represents an update matrix and consists of the inner products of the E basis functions with the B basis functions. The B and E fields are only offset in the direction of differentiation of the E field. Thus, m represents the position of all of the neighboring cells that the E and B fields overlap (depending on the domain of the chosen basis function).

The $\mathbf{U}$ matrices are $2^{3+r_{\max,x}+r_{\max,y}+r_{\max,z}} \times 2^{3+r_{\max,x}+r_{\max,y}+r_{\max,z}}$ in size and can be computed before simulation begins. They take the form

$$
\mathbf{U}^{F_1}_{F_2,m} = \begin{bmatrix}
\left\langle {}_{F_1}\Gamma_1, \dfrac{\partial}{\partial n}\, {}_{F_2}\Gamma_1\big|_m \right\rangle & \left\langle {}_{F_1}\Gamma_1, \dfrac{\partial}{\partial n}\, {}_{F_2}\Gamma_2\big|_m \right\rangle & \cdots & \left\langle {}_{F_1}\Gamma_1, \dfrac{\partial}{\partial n}\, {}_{F_2}\Gamma_L\big|_m \right\rangle \\[2ex]
\left\langle {}_{F_1}\Gamma_2, \dfrac{\partial}{\partial n}\, {}_{F_2}\Gamma_L\big|_m \right\rangle & \left\langle {}_{F_1}\Gamma_2, \dfrac{\partial}{\partial n}\, {}_{F_2}\Gamma_2\big|_m \right\rangle & & \\[2ex]
\vdots & & \ddots & \\[2ex]
\left\langle {}_{F_1}\Gamma_L, \dfrac{\partial}{\partial n}\, {}_{F_2}\Gamma_L\big|_m \right\rangle & & & \left\langle {}_{F_1}\Gamma_L, \dfrac{\partial}{\partial n}\, {}_{F_2}\Gamma_L\big|_m \right\rangle
\end{bmatrix},
$$

$$(51)$$

where F_1 is the field whose update is being found and F_2 is the field on which the update depends. In this case, ${}_{F_1}\Gamma_l$ is the lth row of Γ for the F_1 fields and ${}_{F_2}\Gamma_l\big|_m$ is the lth row of Γ for the F_2 fields, offset by m in the direction of differentiation. The $\partial/\partial n$ term represents the space derivative, in this case $n = y$ or z. If the basis functions used for each field were collocated, (51) would be diagonal. However, because of the offset scheme used, the functions are only collocated in two of the three Cartesian directions. For example, the B_x and E_y fields have the same location in the x and y directions but are offset in the z direction.

If the field offset in the x, y, and z directions is s_x, s_y, s_z, respectively, then the $(2, 2)$ entry of (51) for $\mathbf{E}_{y,i,j,k}$ (same cell as the $\mathbf{B}$ vector, $m = 0$) is

$$
\mathbf{U}^{B_{x1}}_{E_y,m,2,2} = \left\langle \psi^0_{i,0}(x)\varphi_{j+s_y}(y)\varphi_{k+s_z}(z),\ \frac{\partial}{\partial z}\psi^0_{i,0}(x)\varphi_{j+s_y}(y)\varphi_k(z) \right\rangle
$$
$$
= \iiint \psi^0_{i,0}(x)\varphi_{j+s_y}(y)\varphi_{k+s_z}(z)\frac{\partial}{\partial z}\left(\psi^0_{i,0}(x)\varphi_{j+s_y}(y)\varphi_k(z)\right)\partial x\partial y\partial z.
$$

$$(52)$$

This integral can be separated by direction

$$
\iiint \psi_{i,0}^0(x)\varphi_{j+s_y}(y)\varphi_{k+s_z}(z)\frac{\partial}{\partial z}\left(\psi_{i,0}^0(x)\varphi_{j+s_y}(y)\varphi_k(z)\right)\partial x\partial y\partial z
$$
$$
= \int \psi_{i,0}^0(x)\psi_{i,0}^0(x)\partial x \int \varphi_{j+s_y}(y)\varphi_{j+s_y}(y)\partial y \int \varphi_{k+s_z}(z)\frac{\partial\varphi_k(z)}{\partial z}\partial z. \tag{53}
$$

Next, the orthogonality between the collocated basis functions can be exploited,

$$
\int \psi_{i,0}^0(x)\psi_{i,0}^0(x)\partial x \int \varphi_{j+s_y}(y)\varphi_{j+s_y}(y)\partial y \int \varphi_{k+s_z}(z)\varphi_k(z)\partial z
$$
$$
= \Delta x \cdot \Delta y \cdot \int \varphi_{k+s_z}(z)\frac{\partial\varphi_k(z)}{\partial z}\partial z, \tag{54}
$$

and, finally, only a single 1-D integral must be evaluated. Depending on the functions involved, this can be done analytically or numerically. For either case, it can be tabulated, and does not have to be performed for each simulation. It should also be noted that due to the orthogonality of the basis functions, the majority of the entries in (51) are zero.

The procedure shown above can be applied to the remainder of (22)–(26), yielding update equations for all B and D fields. The remaining step in determining the MRTD update scheme is to find the updates for E and H fields from the D and B fields, respectively. As these fields are collocated in space and time, the procedure is less complicated, although potentially more computationally intensive.

2.2.3 Media Discretization

In a general structure, the material properties can vary as a function of position. For this work, only linear, isotropic, nondispersive media are considered. The relationship between the fields in this case is given in (15) and (16). Time localization in this case can be performed by taking the inner product with the time yielding

$$
E_d(x, y, z) = \frac{1}{\varepsilon(x, y, z)}D_d(x, y, z), \tag{55}
$$

$$
H_d(x, y, z) = \frac{1}{\mu(x, y, z)}B_d(x, y, z) \quad d = x, y, \text{ or } z. \tag{56}
$$

The update for each component of E and H can be determined by taking the inner product with the scaling functions. However, because the offset of each field component is different, the update for each component is different. The update of a general E component takes the form

$$
_n\mathbf{E}_{d,i,j,k} = \sum_a \sum_b \sum_c \mathbf{U}^{E_d}_{D_d,i,j,k,a,b,c} \, _n\mathbf{D}_{d,i+a,j+b,k+c}, \tag{57}
$$

where a, b, and d represent the relative positions of the surrounding D basis functions that overlap the $(i, j, k)E$ basis function. The number of entries in these sums is dependent on the basis functions. The coefficients of (57) take the form

$$
U^{E_d}_{D_d,a,b,c\,ij} = \iiint \frac{1}{\varepsilon(x,y,z)} \,{}_{E_d}\Gamma_i \, {}_{D_d}\Gamma_j\Big|_{a,b,c} \, \partial x \partial y \partial z. \tag{58}
$$

Unlike the $\mathbf{U}$ matricies that are used in the discretization of Faraday's and Ampere's laws, most of the entries in these update matricies are nonzero. Furthermore, the matrices vary by position. However, for regions of uniform media, they are diagonal for $l = m = n = 0$, and zero elsewhere.

These equations can be used together in a time-stepping scheme with the discretizations of Ampere's and Faraday's laws to create a general time marching MRTD scheme:

1. Determine B fields from E fields.

2. Determine H fields from B fields.

3. Determine D fields from H fields.

4. Determine E fields from D fields.

5. Repeat until simulation is complete.

2.2.4 Numerical Stability

The scheme that has been presented allows MRTD update schemes to be determined for any basis function. These schemes are fully explicit, the matrix expressions demonstrated only involve multiplication, and all numerical integration is performed before the start of simulation. For the majority of cases, a library of the update equations, **U**, can be constructed, so that the costly numerical integration does not have to be performed before each simulation. Most of the update matrices are very sparse, and a number of clever coding techniques and libraries can be used to reduce the processor and memory load of implementing these updates. Any basis function can be used by calculating the **U** matricies, which can be accomplished by solving a number of integrals, either analytically, or, more likely, numerically. To this point, however, no guidance has been given on choosing the time and space step values. To properly choose these values, the stability and dispersion of the scheme must be taken into consideration.

It is a known limitation of explicit techniques that the time step must limited as a function of the space step to maintain stability. If the time step is not below this limit, the resulting scheme quickly grows without bounds. One notable exception to this is the ADI-MRTD method [10], which is a semiimplicit scheme that remains stable (but not accurate) for any time step. This technique separates the time step into two substeps, and, as it does not alter the spatial grid, it is very similar to ADI-FDTD [11]. The discussion will focus on the traditional MRTD technique. A brief stability analysis of the MRTD technique is presented in [2], while additional studies are presented in [9, 12].

The stability condition for MRTD depends on the choice of the wavelet basis. The general dispersion analysis, however, is performed in the same manner as traditional FDTD techniques [13]. The time update portion (the discretization of the time derivative) and the space update portion (the discretization of the spatial derivative) are split into two separate problems. For numerical stability, the eigenvalues of the spatial problem must contain the eigenvalues of the temporal problem. For FDTD, this results in the condition [13]:

$$\Delta t \leq \frac{1}{\sqrt{\frac{1}{\mu\varepsilon}\left(\frac{1}{\Delta x^2} + \frac{1}{\Delta y^2} + \frac{1}{\Delta z^2}\right)}}. \tag{59}$$

For Battle–Lemarie S-MRTD (scaling function only) [12]

$$\Delta t \leq \frac{1}{\displaystyle\sum_{i=0}^{n_a-1} |a_i| \sqrt{\frac{1}{\mu\varepsilon}\left(\frac{1}{\Delta x^2} + \frac{1}{\Delta y^2} + \frac{1}{\Delta z^2}\right)}}, \tag{60}$$

where

$$s\text{MRTD} = \sum_{i=0}^{n_a-1} |a_i|, \tag{61}$$

the MRTD stability factor is the sum of the magnitudes of the inner products of the offset scaling functions from the spatial update equations (equivalent to the entries in the 1×1 **U** matrices using the notation presented here). In this equation, n_a is the *stencil size*, or the domain of the basis function in cells. The Battle–Lemarie basis functions are symmetric. Alternatively, the sum could be expressed as

$$s\text{MRTD} = \sum_{i=-n_b}^{n_a} |a_i|, \tag{62}$$

where n_a and n_b represent the stencil size of the function in both the positive and negative directions, respectively. In fact, if this expression is used, the stability condition is valid for any basis function. For the Battle–Lemarie scheme [12],

$$s\text{MRTD} = \frac{1}{\displaystyle\sum_{i=0}^{n_a-1} |a_i|} = 0.6371, \tag{63}$$

and the time step for the MRTD scheme is *smaller* than the time step for and FDTD grid with the same cell size. However, as will be shown next, the cells used in Battle–Lemarie MRTD are generally larger than those used in FDTD, and thus the technique is more efficient overall. It is demonstrated in [9] that, for any MRTD wavelet basis, the resolution of the scheme doubles for each addition of a level of wavelet resolution. This is equivalent to saying the time step of a scheme with one level of wavelet resolution in each direction will have a time step one-half of the value of S-MRTD. The resulting stability condition, valid for any MRTD basis, is

$$\Delta t \leq \cfrac{1}{\sum_{i=-n_b}^{n_a} |a_i| \sqrt{\cfrac{1}{\mu\varepsilon}\left(\cfrac{1}{\left(\cfrac{\Delta x}{2^{r_{\max,x}+1}}\right)^2} + \cfrac{1}{\left(\cfrac{\Delta y}{2^{r_{\max,y}+1}}\right)^2} + \cfrac{1}{\left(\cfrac{\Delta z}{2^{r_{\max,z}+1}}\right)^2}\right)}}. \tag{64}$$

The Haar MRTD technique can be analyzed using the same method. Several authors [3, 7–9] have noted the equivalence between Haar S-MRTD and FDTD, and indeed this relationship can be exploited to directly derive the general Haar MRTD stability requirement. Using this equivalence, and applying the resolution doubling argument to (59), the stability for Haar MRTD at any resolution level is [9]

$$\Delta t \leq \cfrac{1}{\sqrt{\cfrac{1}{\mu\varepsilon}\left(\cfrac{1}{\left(\cfrac{\Delta x}{2^{r_{\max,x}+1}}\right)^2} + \cfrac{1}{\left(\cfrac{\Delta y}{2^{r_{\max,y}+1}}\right)^2} + \cfrac{1}{\left(\cfrac{\Delta z}{2^{r_{\max,z}+1}}\right)^2}\right)}}. \tag{65}$$

This is a special case of (64), for the Haar scheme

$$s_{\text{MRTD}} = 1. \tag{66}$$

2.2.5   Numerical Dispersion

Dispersion is defined as the variation of the wavelength, λ, with frequency, f. More commonly, it is discussed as the variation of the wavenumber, k, with angular frequency, ω, where

$$\omega = 2\pi f, \tag{67}$$

$$k = \frac{2\pi}{\lambda}. \tag{68}$$

By substituting the solution for a monochromatic plane wave into Maxwell's equations (assuming a linear, isotropic, nondispersive medium), the following dispersion relationship results:

$$k = \pm \frac{\omega}{c}, \tag{69}$$

where

$$c = \frac{1}{\sqrt{\mu \varepsilon}}. \tag{70}$$

In a 3-D representation, a wavevector, $\mathbf{k}$, is defined, where, in Cartesian coordinates,

$$\mathbf{k} = k_x \hat{i} + k_y \hat{j} + k_z \hat{k} \tag{71}$$

and

$$k = \sqrt{k_x^2 + k_y^2 + k_z^2}. \tag{72}$$

Using these parameters, a phase velocity, v_p, and group velocity, v_g, can be defined,

$$v_p = \pm \frac{\omega}{k} = \pm c, \tag{73}$$

$$v_g = \pm \frac{\partial \omega}{\partial k} = \pm c. \tag{74}$$

These parameters demonstrate that the wavelength and frequency have linear relationship, and that the phase and group velocities are independent of frequency.

These relationships are more complicated in a numerical scheme. A numerical scheme, by its nature, discretizes space into small, but finite, cells. In these cells, the waves cannot propagate in any direction, but rather in the directions defined by the grid. In addition, time is not continuous, but set as a multiple of a discrete time step. With these limitations it should not be surprising that the numerical wave propagation velocity can depend on direction, and, because of the discrete spatial and time steps, frequency. As it has been shown, Haar S-MRTD is equivalent to FDTD, and, as one of the simplest MRTD schemes, it will be used as the first example. For S-MRTD (and FDTD) [14],

$$\left[\frac{1}{c\,\Delta t} \sin\left(\frac{\omega \Delta t}{2} \right) \right]^2 = \left[\frac{1}{\Delta x} \sin\left(\frac{k_x \Delta x}{2} \right) \right]^2 + \left[\frac{1}{\Delta y} \sin\left(\frac{k_y \Delta y}{2} \right) \right]^2$$

$$+ \left[\frac{1}{\Delta x} \sin\left(\frac{k_z \Delta z}{2} \right) \right]^2. \tag{75}$$

For numerical stability, the argument was advanced that increasing the resolution by one level effectively doubled the resolution of the scheme. This was represented in the time stability condition by dividing the space step by $2^{r_{max}+1}$. For the dispersion analysis, this condition holds as well, thus [9],

$$\left[\frac{1}{c\,\Delta t}\sin\left(\frac{\omega\Delta t}{2}\right)\right]^2 = \left[\frac{2^{r_{max,x}+1}}{\Delta x}\sin\left(\frac{k_x\Delta x}{2^{r_{max,x}+2}}\right)\right]^2 + \left[\frac{2^{r_{max,y}+1}}{\Delta y}\sin\left(\frac{k_y\Delta y}{2^{r_{max,y}+2}}\right)\right]^2$$
$$+ \left[\frac{2^{r_{max,z}+1}}{\Delta x}\sin\left(\frac{k_z\Delta z}{2^{r_{max,z}+2}}\right)\right]^2. \tag{76}$$

A similar dispersion analysis was performed in [12] for Battle–Lemarie S-MRTD, and, when made general for any wavelet basis and resolution level,

$$\left[\frac{1}{c\,\Delta t}\sin\left(\frac{\omega\Delta t}{2}\right)\right]^2 = \left[\frac{2^{r_{max,x}+1}}{\Delta x}\left(\sum_{i=n_b}^{n_a-1} a_i \sin\left(\frac{k_x\left(i+1/2\right)\Delta x}{2^{r_{max,x}+2}}\right)\right)\right]^2$$
$$+ \left[\frac{2^{r_{max,y}+1}}{\Delta y}\left(\sum_{i=n_b}^{n_a-1} a_i \sin\left(\frac{k_y\left(i+1/2\right)\Delta y}{2^{r_{max,y}+2}}\right)\right)\right]^2. \tag{77}$$
$$+ \left[\frac{2^{r_{max,z}+1}}{\Delta z}\left(\sum_{i=n_b}^{n_a-1} a_i \sin\left(\frac{k_x\left(z+1/2\right)\Delta z}{2^{r_{max,z}+2}}\right)\right)\right]^2$$

This is the general dispersion relationship for MRTD. As (65) is a special case of (64), (76) is a special case of (77).

It is useful to express the wavenumber in terms of angular frequency. As (77) is a sum of sinusoids, an analytical solution cannot be found for the general case. It can, however, be solved numerically for specific wavelet bases. Several studies have been performed using these equations [4, 9, 12]. For reference, the solution for the wavenumber as a function of angular frequency for Haar MRTD in one of the grid major directions (along an axis), assuming a uniform grid size Δ, is [14]

$$k = \frac{2}{\Delta}\sin^{-1}\left(\frac{1}{S}\sin\left(\frac{\pi S}{N_\lambda}\right)\right), \tag{78}$$

and the phase velocity is

$$v_p = \frac{\pi}{N_\lambda \sin^{-1}\left(\dfrac{1}{S}\sin\left(\dfrac{\pi S}{N_\lambda}\right)\right)} c,$$

(79)

using definitions

$$N_\lambda = \frac{\lambda_\circ}{\Delta} = \frac{c}{\omega \Delta},$$

(80)

$$S = \frac{c\,\Delta t}{\Delta}.$$

(81)

Using this expression, the error between the numerical phase velocity and c can be determined, as well as what values of N_λ (the number of cells per wavelength) yield stable (nondamped) results. This equation is also valid for any wavelet resolution if Δ is the spacing between equivalent grid points. To keep phase error low, it is usual for these schemes to use more than 10 equivalent cells per smallest excited wavelength.

This condition is one of the major arguments for using the MRTD technique. Although it was demonstrated for Haar MRTD that the dispersion relationship, and thus the number of cells needed per wavelength, is the same for the MRTD scheme as in FDTD, other basis functions can use significantly fewer cells per wavelength. It was reported in [12] that a discretization of three to four cells per wavelength using Battle–Lemarie MRTD gave results comparable to FDTD with 10 cells per wavelength.

2.3   IMAGE THEORY FOR PEC MODELING

The techniques that have been presented thus far allow the modeling of arbitrary variation of the permeability and permittivity of the media under simulation. However, most structures of interest to microwave designers also contain metals. For many simulations in the time domain, these metals are treated as PECs, and this limitation will be used in this case as well. The treatment of a metal as a PEC is necessary, because otherwise the metal itself must be simulated (resulting in significantly increased number of grid points) or frequency dependent loss characteristics must be applied to the PEC.

In the prototypical time-domain technique, FDTD, PEC structures are simulated in a simple and straightforward manner. The PECs are located along grid boundaries, and the electric field values that are tangential to the PEC structures are zeroed each time step (after their updates are calculated). This explicitly enforces (17). The normal coefficients do not require any special processing, because they are offset from the PEC location. Because of the extended, and often overlapping, nature of MRTD basis functions, this simple processing is not possible in MRTD. Instead, PEC boundary conditions are traditionally applied to MRTD codes through image theory.

Several papers have been published that discuss the implementation of image theory to the modeling of PEC structures in MRTD [2, 9, 12, 15–17]. Using this method, PECs are modeled through the introduction of artificial image scaling and wavelet coefficients on the opposite side of the PEC. Tangential E field components and the normal H component are opposite their components across the image, while other components are identical. In this manner, when the real and image wavelet and scaling coefficients are summed at the PEC interface, the result is zero, the PEC boundary condition. This technique is difficult to apply in a general 3-D code, because a large amount of bookkeeping is required for each PEC in the grid. For any wavelet that contains a PEC within its stencil, a unique update scheme must be applied. In addition, multiple images, caused by the close proximity of multiple PECs, can exacerbate this problem further. While it has been stated in several publications that this technique can be applied to complex structures, no publications have been made detailing its use for a general 3-D structure with multiple PECs. At maximum, external boundaries have been presented.

A 1-D example of the application of image theory is presented in Fig. 7. This example uses biorthogonal scaling functions [5]. These functions are used because they have a relatively compact stencil size of four cells. The shaded area represents the width of the stencil. All magnetic field points within this distance of

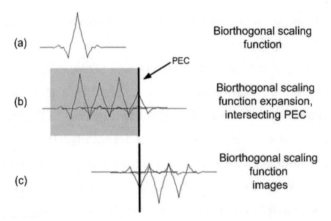

FIGURE 7: Image theory in one dimension, using biorthogonal scaling functions.

the PEC must use the images of the electric field on the opposite of the PEC when computing an update. If this PEC intersects the grid (the simulation is performed on both sides of the PEC), the actual field values on the opposite of the PEC are not used (the PEC effectively decouples these fields). In addition, when the fields are reconstructed, the images must be used in the reconstruction algorithm. This same method can be used to simulate perfect magnetic conductors (PMC) by imaging the magnetic fields across the boundary.

When performing simulations, using any wavelet basis other that the Haar basis functions, this treatment is necessary to terminate the space. In order to perform the leapfrog update scheme used in MRTD, the fields must be known on the outer boundary. Using image theory, the fields outside the computational space are computed as images of the internal fields. These field are not updated, and as they are the same (but different in magnitude) as the internal fields, they do not require additional memory. If the span of the basis functions extends more that the size of the computational domain across the boundary, additional images can be used. This has been termed the multiple image technique (MIT) [17].

2.4   FDTD

It has already been stated that FDTD is often used as a benchmark for other time domain schemes, such as MRTD. The chief advantage of the FDTD method is

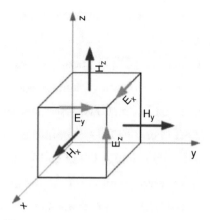

FIGURE 8: Yee cell in 3D.

its simplicity; it was originally derived using central differences with Maxwell's equations. However, the key aspect of the FDTD method does not come from its relatively simple derivation, but the arrangement of its grid points.

FDTD was originally developed in 1967 by Yee [1]. The feature of FDTD that makes it both simple and widely applicable is the Yee cell, which is presented in Fig. 8. A simple Cartesian grid is used, with the fields offset from the grid points in a specific manner. In Fig. 8 the electric field components are offset from the grid points half a cell in their coordinate direction (E_x in x, for example) and the magnetic fields are offset in the directions normal to their field components.

Using this simple field arrangement, and by offsetting the magnetic and electric fields by half a step in time, central differences can be applied to Faraday's and Ampere's law, creating a system of equations that can be used in a time marching scheme to update the fields. For example, the FDTD discretization of (21) is

$$
{}^{n+1/2}H_{x,i,j,k} = {}_{n-1/2}H_{x,i,j,k} + \frac{\Delta t}{\varepsilon}\left[\frac{{}_nE_{y,i,j,k} - {}_nE_{y,i,j,k-1}}{\Delta z} - \frac{{}_nE_{z,i,j,k} - {}_nE_{z,i,j-1,k}}{\Delta y}\right], \quad (82)
$$

if the convention that has been used in the MRTD case is applied, where all field components are offset in the positive direction, and the offsets are not explicitly shown in the field components ($E_{y,i,j,k}$ is located at $i\Delta x$, $(j + 1/2)\Delta y$, $k\Delta z$, for example). In this equation, the constitutive relationship (15) is directly applied, as the updates are for actual field values at points instead of scaling/wavelet coefficients that span a much wider area.

It is useful to note that the FDTD update equations are exactly the same as Haar S-MRTD (scaling function only) update equations. The method-of-moments, applied with pulse basis functions, offset according to the scheme that has been presented for MRTD and FDTD, is equivalent to performing a central difference approximation. Furthermore, FDTD and Haar MRTD are equivalent for any r_{max} if all wavelets are used for all points in the grid. This can be demonstrated by transforming the MRTD scaling/wavelet scheme into a pointwise (where the actual field values are used) scheme, [18], or by noting that the Haar MRTD dispersion relationship for an arbitrary resolution level is the same as FDTD, if the FDTD cells are the same size as the MRTD equivalent grid points.

C H A P T E R   3

Modeling of Practical Structures

The MRTD scheme presented in the previous chapter provides a framework for applying multiresolution principles to the simulation of modern microwave structures. Every paper published on MRTD discusses two major advantages of the MRTD technique:

1. The basis functions employed in the MRTD technique allow fewer cells to be used per maximum excited wavelength (compared to FDTD), allowing a more efficient overall simulation.

2. Wavelets can be added and subtracted dynamically during simulation, resulting in an adaptive scheme that automatically tailors its computational requirements to the complexity of propagating signals.

The number of cells required per wavelength is discussed in the previous chapter. The dispersion relationship, (77), can be solved numerically for any wavelet basis. It is shown in several papers [2, 4, 9, 20] that a variety of wavelet bases can be employed that allow less dense grids to be used. There are two distinct costs of applying these techniques. The first is determining the update coefficients, such as those used in (50) and (57), that result from Ampere's and Faraday's laws. However, these coefficients can be calculated before the start of simulation, and it is possible to build a library of the coefficients for several common cases. The second cost of applying the technique is the field updates themselves, which, due to the stencil of the basis functions, involve several surrounding field coefficients. Thus, while fewer cells are required per wavelength when compared to FDTD, more operations are required to calculate each field update. A comprehensive comparison of the benefits from using fewer cells in MRTD compared to the more simple updates used in FDTD has not been presented in literature.

Another difficulty of using general wavelet bases is that of simulating lossy media. The derivations of MRTD update equations presented in the previous chapter neglected loss. When Ohmic losses are added (considering both electric and magnetic loss), Faraday's and Ampere's laws become

$$\nabla \times \mathbf{E}(t) = -\frac{\partial \mathbf{B}(t)}{\partial t} - \sigma^* \mathbf{B}(t) \tag{83}$$

$$\nabla \times \mathbf{H}(t) = \frac{\partial \mathbf{D}(t)}{\partial t} + \sigma \mathbf{D}(t). \tag{84}$$

The procedure for determining the MRTD update procedures described in the previous chapter can be applied in this case as well. For example, the update for the B_x field is determined from

$$\frac{\partial B_x}{\partial t} = \frac{\partial E_y}{\partial z} - \frac{\partial E_z}{\partial y} - \sigma^* B_x. \tag{85}$$

To determine the time localized form of this equation, the inner product of both sides of (85) must be taken with $h_{n+1/2}(t)$. This process is slightly different than the case without loss, because both the time derivative of the B field and the B field itself are used in this equation. In Fig. 9, it is noted that the pulse used in the time localization overlaps the B field from two time steps. The inner product

$$\langle h_n(t), B_x \rangle = \frac{^{n+1/2} B_x(\mathbf{r}) + {}_{n-1/2} B_x(\mathbf{r})}{2} \Delta t \tag{86}$$

therefore includes terms from both the time step being updated and the previous time step. This is a statement of the semiimplicit approximation used in FDTD [14], but it is derived in this case as a direct result of the chosen basis functions.

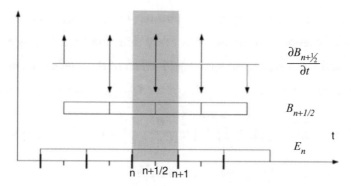

FIGURE 9: Time localization (pulse basis functions) for lossy case.

Using (86), the time localized form of (85) is

$$_{n+1/2}B_x(\mathbf{r}) = {}_{n-1/2}B_x(\mathbf{r}) - \sigma^*\Delta t \frac{_{n+1/2}B_x(\mathbf{r}) + {}_{n-1/2}B_x(\mathbf{r})}{2}$$

$$+ \Delta t\left[\frac{\partial_n E_y(\mathbf{r})}{\partial z} - \frac{\partial_n E_z(\mathbf{r})}{\partial y}\right]. \tag{87}$$

Collecting terms

$$\left(1 + \frac{\sigma^*\Delta t}{2}\right)_{n+1/2}B_x(\mathbf{r}) = \left(1 - \frac{\sigma^*\Delta t}{2}\right)_{n-1/2}B_x(\mathbf{r})$$

$$+\Delta t\left[\frac{\partial_n E_y(\mathbf{r})}{\partial z} - \frac{\partial_n E_z(\mathbf{r})}{\partial y}\right] \tag{88}$$

and finally solving for $_{n+1/2}B_x(\mathbf{r})$

$$_{n+1/2}B_x(\mathbf{r}) = \left(\frac{2 - \sigma^*\Delta t}{2 + \sigma^*\Delta t}\right)_{n-1/2}B_x(\mathbf{r})$$

$$+ \left(\frac{2\Delta t}{2 + \sigma^*\Delta t}\right)\left[\frac{\partial_n E_y(\mathbf{r})}{\partial z} - \frac{\partial_n E_z(\mathbf{r})}{\partial y}\right]. \tag{89}$$

When the space localization is performed to determine the updates for the individual wavelet coefficients, the following results:

$$_{n+1/2}\mathbf{B}_{x,i,j,k} = \sum_a\sum_b\sum_c \mathbf{U}^{B_x}_{B_x,a,b,c\,n-1/2}\mathbf{B}_{x,i+a,j+b,k+c}$$

$$+\frac{\Delta t}{\Delta x\Delta y\Delta z}\left[\begin{array}{l}\sum_a\sum_b\sum_c \mathbf{U}^{B_x}_{E_y,a,b,c\,n-1/2}\mathbf{E}_{y,i+a,j+b,k+c}\\[4pt]+\sum_a\sum_b\sum_c \mathbf{U}^{B_x}_{E_z,a,b,c\,n-1/2}\mathbf{E}_{z,i+a,j+b,k+c}\end{array}\right], \tag{90}$$

using the notation presented in the last chapter. In this case, however, it is noted that the basis functions can be offset in every direction (denoted by a, b, and c, in the x, y, and z directions, respectively), instead of the single direction that was denoted by m in Chapter 2. The entries of the $\mathbf{U}^{B_x}_{B_x,a,b,c}$ matrices take the form

$$U_{ij} = \left\langle {}_{B_x}\Gamma_i, \sigma^* {}_{B_x}\Gamma_j\big|_{a,b,c} \right\rangle = \iiint \left(\frac{2 - \sigma^*\Delta t}{2 + \sigma^*\Delta t} \right) {}_{B_x}\Gamma_i {}_{B_x}\Gamma_j\big|_{a,b,c} \partial x \partial y \partial z, \quad (91)$$

and the entries of the $\mathbf{U}^{B_x}_{E_y,a,b,c}$ matrices take the form

$$U_{ij} = \left\langle {}_{B_x}\Gamma_i, \sigma^* {}_E\Gamma_j\big|_{a,b,c} \right\rangle = \iiint \left(\frac{2\Delta t}{2 + \sigma^*\Delta t} \right) {}_B\Gamma_i \frac{\partial}{\partial z} {}_E\Gamma_j\big|_{a,b,c} \partial x \partial y \partial z, \quad (92)$$

where $\Gamma_j\big|_{a,b,c}$ denotes the scaling/wavelet function offset by a, b, and c (cells) in the x, y, z directions, respectively, from the coefficient update being calculated. In the general case of $\sigma^*(x, y, z)$ varying constantly in each direction, the $\mathbf{B_x}$ coefficients depend on surrounding $\mathbf{B_x}$ coefficients from the previous time step. This complicates the update of the coefficients, as when implemented on a computer, the array of updated values must be kept separate from previous values. In the lossless scheme, the fields being updated depend only on their previous value, and the updated fields can be stored in the same array as the previous fields. This doubles the amount of memory required to perform the scheme. In addition, it significantly increases the computational burden.

In the lossless case, the updates of any one coefficient depend on the previous value of the coefficient and the values of two other fields (the normal E fields in the B update case, and the normal H fields in the D case). For these fields, a sum of coefficients (multiplied by the entries in the $\mathbf{U}$ matrix) from neighboring grid points is required, and the number of grid points depends on the stencil of the basis function. For example, a stencil of 12 (6 in the positive, 6 in the negative direction) cells is normally used for Battle–Lemarie wavelets [16]. However, because the fields are offset only in one direction, the sum must only be performed in one direction (as in (50)). In the case including loss, however, the addition of σ into the inner products destroys the orthogonality in the inner product integrals, and the sums must be performed in all three directions. For a stencil size n, the result is a summation over n^3 elements. As an example, in the lossless case, for Battle–Lemarie wavelets using a stencil size of 12, the total number of neighboring cells required for the update is

Previous value + normal field 1 + normal field 2

$$1 + n + n = 2n + 1, \tag{93}$$
$$1 + 12 + 12 = 25, \tag{94}$$

while for the lossy case

$$n^3 + n^3 + n^3 = 3n^3, \tag{95}$$
$$1728 + 1728 + 1728 = 5184. \tag{96}$$

This significantly complicates the method. However, for stencil sizes of 1 (such as Haar basis functions) the requirements are the same in the lossy and lossless case. In addition, there are several classes of wavelets and wavelet like functions that allow this condition to be reduced. One example is the Daubechies' wavelets with vanishing moments [19, 20].

The final difficulty of simulating complex structures using general wavelet bases is that of representing PEC structures within an MRTD grid. In the previous chapter image theory was discussed; the primary example presented, as well as the examples shown in the referenced literature, was for the case of a PEC wall terminating the structure. A general method for simulating arbitrarily placed internal PEC structures has not been presented in literature. This makes simulation of most microwave structures difficult. One notable exception is [20] which utilizes an FDTD/Haar-MRTD interface to simulate highly detailed structure in FDTD, and MRTD to simulate open areas. However, this technique cannot be readily expanded to other wavelet bases. In this chapter a method is presented that allows arbitrarily placed PEC structures to be simulated; however, it is shown that it is practical only for Haar-MRTD schemes.

As this monograph is an introduction to the MRTD method, and Haar wavelets are used to keep the examples relatively simple, in this chapter a technique is presented that allows subcell structures to be modeled within Haar MRTD cells. This technique is shown to be a bridge between pointwise field updates, such as those used in FDTD and the wavelet/scaling updates used in MRTD. This technique allows the adaptive resolution characteristics of MRTD to be exploited for the simulation of any structure.

3.1   GENERAL, EXPLICIT, SUBCELL FIELD MODIFICATION

One challenge in MRTD modeling is representing arbitrary structures in such a way that the time-space adaptive grid can be used effectively. To accomplish this, it is necessary to determine a method of representing PEC structures that are smaller than an MRTD cell. Using the Haar basis functions, it is possible to apply the PEC boundary condition in a similar method as FDTD grids. It is shown that the method effectively provides a bridge between pointwise electromagnetic effects and the distributed wavelet field representation used in MRTD.

In FDTD, which is equivalent to Haar S-MRTD, PEC structures are explicitly represented by zeroing electric field values that are tangential to PEC field locations. The update equations for FDTD can be determined in the same manner as the MRTD equations in the previous chapter. In fact, the B_x update equation is a special case of (50). Similarly, the D_x update equation is [14]

$$_{n+1}D_{x,i,j,k} = {}_nD_{x,i,j,k} + \Delta t \left[\frac{_{n+1/2}H_{z,i,j,k} - {}_{n+1/2}H_{z,i,j-1,k}}{\Delta y} - \frac{_{n+1/2}H_{y,i,j,k} - {}_{n+1/2}H_{y,i,j,k-1}}{\Delta z} \right], \tag{97}$$

in the lossless case (for the purpose of applying PEC boundary conditions, loss is irrelevant). In FDTD, the constitutive relationships are applied at each cell, so

$$_{n+1}E_{x,i,j,k} = {}_nE_{x,i,j,k} + \frac{\Delta t}{\varepsilon} \left[\frac{_{n+1/2}H_{z,i,j,k} - {}_{n+1/2}H_{z,i,j-1,k}}{\Delta y} - \frac{_{n+1/2}H_{y,i,j,k} - {}_{n+1/2}H_{y,i,j,k-1}}{\Delta z} \right]. \tag{98}$$

Note that in this notation, the spatial offsets from the grid points are not written explicitly, similar to the MRTD case in the previous chapter (but follow the same scheme as MRTD). In this case, the electric fields are located half a cell from point $i\Delta x, j\Delta y, k\Delta z$ in the direction of their field component, and the magnetic fields are located half a cell from $i\Delta x, j\Delta y, k\Delta z$ in the two directions normal to the field component that they represent. A 2-D cross-section of an FDTD grid intersected by a PEC is presented in Fig. 10.

In FDTD, the PEC is placed along the locations of the electric field points. The practical result is that the size of the structure being simulated is constrained by the grid. To represent field variation caused by the PECs, several cells are usually placed across each PEC. The PEC condition is applied by first updating the electric fields, and then setting the field values that overlap with the PEC locations to zero. This is possible because the electric field update equations, such as (98), use only previous field values.

The time update scheme used in MRTD takes the same form as FDTD (the same basis functions are used for the time discretization). Instead of using

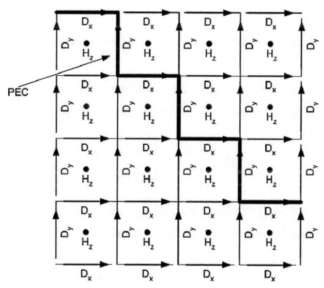

FIGURE 10: FDTD grid intersected by PEC.

image theory, which creates artificial scaling/wavelet coefficients to apply PEC conditions, it would be convenient to explicitly enforce the boundary condition of zero tangential electric field on the MRTD grid.

For a general wavelet basis, using the notation presented in the previous chapter, the electric field is reconstructed as

$$_n E_{\text{dir}}(\mathbf{r}) = \sum_i \sum_j \sum_k \Gamma^T_{i,j,k\,n} \mathbf{E}_{\text{dir},i,j,k}. \tag{99}$$

This function is used to give the fields at every point in the grid. It is useful to note that the MRTD scheme never updates the field values themselves, as in FDTD, but only updates the scaling/wavelet coefficients that then must be reconstructed to determine field values. This is important to note when probing the fields during simulation, as the fields must be reconstructed at points of interest. When probing fields during simulation, (99) is unwieldy; reconstructing the fields for the entire grid for each time step is computationally prohibitive. Instead, the fields for an individual cell can be reconstructed using

$$_n E_{\text{dir},i,j,k}(\mathbf{r}) = \sum_{a=-n_a}^{n_b} \sum_{b=-n_a}^{n_b} \sum_{c=-n_a}^{n_b} \Gamma^T_{i+a,j+b,k+c\,n} \mathbf{E}_{\text{dir},i+a,j+b,k+c}, \tag{100}$$

where n_a and n_b represent the size of the stencil (the overlap of the wavelet/scaling function in cells).

Once the fields are reconstructed, a localized PEC can be applied by multiplying the reconstructed electric field by a function $P(\mathbf{r})$, where

$$P(\mathbf{r}) = \begin{cases} 0 & \text{PEC location} \\ 1 & \text{Elsewhere} \end{cases}. \tag{101}$$

Once the PEC boundary condition has been applied, the fields must be transformed back into the wavelet domain. This can be accomplished by applying a wavelet transform

$$\mathbf{E}_{\text{dir},i,j,k} = \langle \Gamma_{i,j,k}, P(\mathbf{r})_n E_{\text{dir},i,j,k}(\mathbf{r}) \rangle. \tag{102}$$

Of course, the new components must be found at all locations whose stencils include a PEC location. By combining (99) and (102), the wavelet coefficient with the PEC condition applied can be found directly from other scaling/wavelet coefficients using

$$\mathbf{E}_{\mathrm{dir},i,j,k} = \left\langle \Gamma_{i,j,k},\, P(\mathbf{r}) \sum_i \sum_j \sum_k \Gamma^T_{i,j,k\,n} \mathbf{E}_{\mathrm{dir},i,j,k} \right\rangle. \tag{103}$$

In the case of a general wavelet resolution, there are several difficulties in applying this technique. The first is that, for arbitrarily placed PECs, the reconstruction, and subsequent wavelet decomposition, of the entire grid is required for each time step. This procedure likely requires more computation time than the field updates and is therefore not practical. In addition, the implicit assumption in the above procedure is that the fields, when deconstructed back into the wavelet domain after the application of the PEC boundary condition, reconstruct to the same values at all non-PEC locations, and zero at PEC locations. While all of the wavelet bases discussed in this book are complete, this condition is true only for an infinite level of wavelet resolution. For a limited wavelet resolution, which is kept the same before and after the application of the PEC, the application of the PEC as outlined here will result in modified fields outside the PEC region, and nonzero fields inside the PEC region. This is not consistent with Maxwell's equations.

For this technique to be applied successfully, the wavelet basis must satisfy two conditions:

1. The scaling/wavelet functions for one cell must not overlap with a neighboring cell.

2. The application of a PEC boundary condition (zeroing the field) over a range within an MRTD cell must not affect neighboring field values.

The first condition allows the PEC boundary condition to be applied locally. This means that only the scaling/wavelet coefficients in a single cell must be modified to apply the PEC boundary condition. The second ensures that the fields will not undergo nonphysical modifications when the PEC condition is applied. Both of these conditions are satisfied by the Haar wavelet basis.

3.2   PROPERTIES OF WAVELET DISCRETIZATIONS

3.2.1   HAAR Wavelets

To demonstrate that the Haar basis can be used to explicitly apply subcell PEC boundary conditions, it is first useful to study how the Haar basis functions discretize field values. The Haar scaling and wavelet functions are defined in the previous chapter. When the Haar wavelets are reconstructed, they are constantly valued over discrete regions. For example, Fig. 3 presents the 2-D Haar scaling function and wavelets for $r_{\max} = 0$ in both directions. When these functions are summed, they yield four independent regions of constant field value. These regions are centered at the equivalent grid points. As was stated previously, the number of equivalent grid points is the same as the number of basis functions. In the 3-D case, the number of equivalent grid points (and basis functions) is

$$\text{number of equivalent grid points} = 2^{3+r_{\max,x}+r_{\max,y}+r_{\max,z}}, \tag{104}$$

which is equivalent to (28) for the 3-D case.

As another example, the $r_{\max} = 1$, in all directions, wavelets for the 2-D case are presented in Fig. 11. In this case there are $2^{2+r_{\max,x}+r_{\max,y}} = 2^{2+1+1} = 16$ coefficients per cell. In this figure, each of the wavelets has two possible, identical in magnitude, values, with opposite sign. These are represented in the figure by different shading. For the wavelet/wavelet case, these represent four distinct regions. For the highest level wavelet/wavelet terms (the $\psi_p^{r_{\max}}(x)\, \psi_p^{r_{\max}}(y)$ terms), the equivalent grid points are at the center of the areas of constant magnitude. Each of the highest level wavelet/wavelet terms represents four equivalent grid points in the 2-D scheme.

This discussion is easily extended to 3D. In the $r_{\max} = 1$ case an additional four scaling/wavelet coefficients are required to describe the fields in the z direction. The total number of basis functions in this case is $2^{3+r_{\max,x}+r_{\max,y}+r_{\max,z}} = 2^{3+1+1+1} = 64$. In this case, the regions of constant value are rectangular solids, and for the wavelet/wavelet terms there are eight areas of constant value. Again, the center of these regions for the highest resolution wavelet/wavelet terms represents the

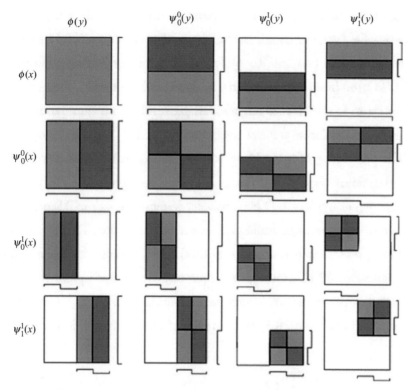

FIGURE 11: Haar scaling and wavelet functions in 2-D, $r_{max} = 1$.

equivalent grid points. When the fields are reconstructed they add to independent values, located at the equivalent grid points. The domain of each value is

$$L = \frac{\Delta D}{2^{1+r_{max,D}}},\qquad(105)$$

in each dimension, D. In the $r_{max} = 1$ case, the value is constant at each equivalent grid point over a range of $\Delta x/4 \times \Delta y/4 \times \Delta z/4$.

One important note is that, for any cell, the fields can be represented in an equivalent manner by scaling functions alone. In this case the scaling functions must be the size of the equivalent grid points. The advantage of using the wavelets is that the resolution can be varied on a cell by cell basis. In cells containing complex structures, high resolution cells can be used. In surrounding cells, lower resolution can be used; Additionally, the wavelet resolution can be varied during simulation. For complex field variation, high resolution wavelets can be used; for less variation, lower resolution can be used. To apply a similar method with scaling functions only (FDTD), an interpolation scheme must be used to interface the high and low resolution areas. In the MRTD scheme, variable resolution is automatically applied by zeroing high resolution wavelet coefficients when they are not needed.

The practicality of using Haar wavelet addition/subtraction as a method of representing fields varying at different rates can be represented analytically and graphically. For example, a 1-D Haar wavelet system with $r_{max} = 0$ is presented. In this scheme there is only one scaling and one wavelet function. These functions can be used to represent any two values. Fig. 12(a) shows the Haar scaling and wavelet function and Fig. 12(b) shows an example of a dual valued piecewise constant function that can be represented using these values. The function $f(x)$ in Fig. 12(b) has value c in the range (0, 0.5) and the value d in the range (0.5, 1). The scaling

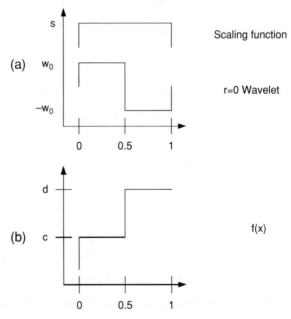

FIGURE 12: (a) Haar scaling and $r_{max} = 0$ wavelet and (b) sample function that can be represented using (a).

function has value s in the range $(0, 1)$ and 0 elsewhere. The wavelet function is valued w_0 in the range $(0, 0.5)$ and $-w_0$ in the range $(0.5, 1)$.

The magnitudes of the scaling/wavelet functions can be found using the system

$$
\begin{aligned}
s + w_0 &= c, \\
s - w_0 &= d.
\end{aligned}
$$ (106)

This system can be easily solved to yield

$$
s = \frac{c + d}{2} \qquad w_0 = \frac{c - d}{2}.
$$ (107)

While this is an extremely basic example it demonstrates an important property of Haar discretizations. The scaling function is the average of the discrete values of $f(x)$. For this two-valued function example, if the values are identical, $w_0 = 0$, and the wavelet coefficient is not needed. In practical simulations, w_0 can be neglected if it is significantly smaller than the scaling function.

This same scheme holds for higher level wavelets. For contrast, a similar example is presented for $r_{max} = 1$. In Fig. 13 the addition of the level one wavelets (the tails are removed for $r_{max} = 1$ to demonstrate that the wavelets represent two independent functions) allows the piecewise constant valued function to have a maximum of four independent values. The wavelets have two values, equal in magnitude and opposite in sign, and for simplicity the magnitude is indicated in the center of each function. In this case a system of four equations can be constructed to determine the scaling/wavelet function magnitudes. The solution of this system,

$$
\begin{aligned}
s &= \frac{c + d + e + f}{4} \qquad w_0 = \frac{c + d - e - f}{4} \\
w_{0,0} &= \frac{c - d}{2} \qquad\qquad w_{0,1} = \frac{e - f}{2}
\end{aligned}
$$ (108)

shows that the scaling function still represents the average of the function. The sum of the scaling term and the 0th level wavelet represents the average on either half of the domain. If the variation of the field on either half of the domain is small, the high-level wavelets can be ignored.

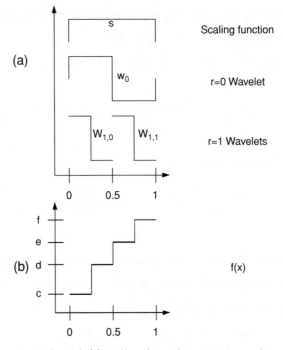

FIGURE 13: Haar example with (a) scaling through $r_{max} = 1$ wavelets and (b) an example function that can be represented with these functions.

One other interesting property of Haar wavelet expansions related to the representation of PEC structures can be demonstrated using this example. The values of the scaling/wavelet functions presented in (108) can be determined regardless of the values of $f(x)$. If $f(x)$ is zeroed over one of the ranges, the remainder of the values can still be represented using the scaling/wavelet basis. The values of all of the basis functions in the cell change, but the values obtained when the functions are summed remain the same.

In the two examples that have been presented of Haar wavelet decompositions, the values of the Haar wavelets were found using a system of linear equations. In the general case of a continuous function, the method of moments can be applied. In the special cases that have been presented of piecewise constant functions with constant domains equal to the equivalent grid point dimensions, the method of moments discretization reduces to a system of linear equations. If the values of the function over these constant areas are represented as a vector $\mathbf{F}$ and the scaling/wavelet function magnitudes are represented as a vector $\mathbf{F_w}$, then they are related by a reconstruction matrix $\mathbf{R}$ where

$$\mathbf{F} = \mathbf{RF_w}. \tag{109}$$

This representation is also valid for the 2-D and 3-D cases. Because there are the same number of equivalent grid points as scaling/wavelet functions, $\mathbf{R}$ is square. In the two and three dimensional cases, any ordering of the wavelet coefficients and equivalent grid points can be used, it only affects the positioning of the coefficients in the $\mathbf{R}$ matrix. $\mathbf{R}$ can be generated quickly by examining the contribution of each wavelet/scaling coefficient to each equivalent grid point.

Once the reconstruction matrix is determined, the wavelet transform can be easily performed for arbitrary values of $\mathbf{F}$. A wavelet transform matrix, $\mathbf{W}$ is defined as

$$\mathbf{W} = \mathbf{R}^{-1} \tag{110}$$

and

$$\mathbf{F_w} = \mathbf{WF}. \tag{111}$$

These matrices provide a quick and easy transition between scaling/wavelet co-efficients and allow the PEC boundary condition to be explicitly applied at any equivalent grid point in the Haar MRTD scheme without affecting neighboring field values. The only restriction is that the metals must be the size of an equivalent grid point. By increasing the resolution to the appropriate level, arbitrary structures can be represented.

3.3   HAAR SUBCELL PEC APPLICATION

The discussion in the previous section demonstrates that Haar wavelets can be used to apply the explicit PEC method that has been presented. The Haar wavelets are nonoverlapping, and thus a modification of the fields in one cell does not affect neighboring cells. In addition, the reconstruction matrix can be used to quickly determine the field values within one cell, which can be subsequently modified and transformed back into the wavelet domain. The MRTD update algorithm that was presented in the previous chapter can be modified (the added steps are indicated in boldface):

1. Determine B fields from E fields.

2. Determine H fields from B fields.

3. Determine D fields from H fields.

4. **Reconstruct D fields in PEC grid locations.**

5. **Zero fields tangential to PECs.**

6. **Transform D fields back to wavelet domain.**

7. Determine E fields from D fields.

8. Repeat until simulation is complete.

The PEC condition is applied directly to the D fields after they are updated. Mathematically, these steps are relatively simple. First, the fields are reconstructed in the cell where the PEC is to be applied,

$$\mathbf{D}_{\mathrm{dir},i,j,k} = \mathbf{R}\mathbf{D}_{\mathbf{w},\mathrm{dir},i,j,k}. \tag{112}$$

In this case the subscripts indicate the fields at cell i, j, k. The vector $\mathbf{D}_{\mathrm{dir},i,j,k}$ contains the reconstructed field values in the cell. The entries in this vector that correspond to positions of PECs are zeroed. This can also be represented mathematically if a matrix $\mathbf{I_P}$ is defined, where $\mathbf{I_P}$ is the identity matrix with the rows that correspond to the PEC locations zeroed. The application of the PEC to the field becomes

$$\mathbf{D}_{\mathbf{PEC},\mathrm{dir},i,j,k} = \mathbf{I_P}\mathbf{D}_{\mathrm{dir},i,j,k}. \tag{113}$$

To continue the MRTD field updates, the $\mathbf{D}$ fields must be transformed back to the wavelet domain

$$\mathbf{D}_{\mathbf{w},\mathrm{dir},i,j,k} = \mathbf{R}^{-1}\mathbf{D}_{\mathbf{PEC},\mathrm{dir},i,j,k}. \tag{114}$$

The procedure that has been presented requires that the $\mathbf{D}$ vector be multiplied by three matrices. These matrices are $2^{3+r_{\max,x}+r_{\max,y}+r_{\max,z}} \times 2^{3+r_{\max,x}+r_{\max,y}+r_{\max,z}}$ in size, for the $r_{\max} = 1$ case the matrix is 64×64, containing 4096 entries. While many of the entries in these matrices are zero, and thus the multiplication can be significantly simplified, this procedure still adds nonnegligible computational overhead. There are two methods that can be used to apply this process in a more efficient manner.

First, the three steps (112)–(114) can be combined. Each step is a matrix multiplication, so an alternate method is

$$\mathbf{D}_{\mathbf{w},\text{dir},i,j,k} = \mathbf{P}_{\text{dir},i,j,k}\mathbf{D}_{\mathbf{w},\text{dir},i,j,k} = \mathbf{R}^{-1}\mathbf{I}_{\mathbf{P}}\mathbf{R}\mathbf{D}_{\mathbf{w},\text{dir},i,j,k}. \tag{115}$$

Before simulation begins, the $\mathbf{P}_{\text{dir},i,j,k}$ matrices can be calculated, and then the PEC condition can be quickly applied with a single matrix multiplication. This method is a simple way to modify an existing Haar MRTD code to add local PEC modeling. An even more efficient method is to directly modify the MRTD update equations.

In Haar MRTD the D field update equations can be calculated using the same procedure as (50), yielding

$$_{n+1}\mathbf{D}_{x,i,j,k} = {}_{n}\mathbf{D}_{x,i,j,k} + \frac{\Delta t}{\Delta x \Delta y \Delta z}\left[\begin{array}{l}\mathbf{U}^{D_x}_{H_y,1\,n+1/2}\mathbf{H}_{y,i,j,k+1} + \mathbf{U}^{D_x}_{H_y,0\,n+1/2}\mathbf{H}_{y,i,j,k} \\ + \mathbf{U}^{D_x}_{H_z,1\,n+1/2}\mathbf{H}_{z,i,j+1,k} + \mathbf{U}^{D_x}_{H_z,0\,n+1/2}\mathbf{H}_{z,i,j,k}\end{array}\right], \tag{116}$$

$$_{n+1}\mathbf{D}_{y,i,j,k} = {}_{n}\mathbf{D}_{y,i,j,k} + \frac{\Delta t}{\Delta x \Delta y \Delta z}\left[\begin{array}{l}\mathbf{U}^{D_y}_{H_x,1\,n+1/2}\mathbf{H}_{x,i,j,k+1} + \mathbf{U}^{D_y}_{H_x,0\,n+1/2}\mathbf{H}_{x,i,j,k} \\ + \mathbf{U}^{D_y}_{H_z,1\,n+1/2}\mathbf{H}_{z,i+1,j,k} + \mathbf{U}^{D_y}_{H_z,0\,n+1/2}\mathbf{H}_{z,i,j,k}\end{array}\right], \tag{117}$$

$$_{n+1}\mathbf{D}_{z,i,j,k} = {}_{n}\mathbf{D}_{z,i,j,k} + \frac{\Delta t}{\Delta x \Delta y \Delta z}\left[\begin{array}{l}\mathbf{U}^{D_z}_{H_x,1\,n+1/2}\mathbf{H}_{x,i,j+1,k} + \mathbf{U}^{D_z}_{H_x,0\,n+1/2}\mathbf{H}_{x,i,j,k} \\ + \mathbf{U}^{D_z}_{H_y,1\,n+1/2}\mathbf{H}_{y,i+1,j,k} + \mathbf{U}^{D_z}_{H_y,0\,n+1/2}\mathbf{H}_{y,i,j,k}\end{array}\right]. \tag{118}$$

In the Haar MRTD case, the sums are not required, because the overlap of the D and B basis functions extends only to the nearest neighbors. All of these equations have the same format; the only differences between the equations are the field values involved and the entries of the $\mathbf{U}$ matrices. It should be noted that the $\mathbf{U}$ matrices are the same size as the reconstruction/wavelet decomposition matrices. Neglecting unneeded subscripts, any of (116)–(118) can be represented as

$$\mathbf{D}_{\mathbf{W}\mathrm{dir}1} = \mathbf{D}_{\mathbf{W}\mathrm{dir}1} + \frac{\Delta t}{\Delta x \Delta y \Delta z} \begin{bmatrix} \mathbf{U}_1 \mathbf{H}_{\mathbf{W}\mathrm{dir}2,1} + \mathbf{U}_2 \mathbf{H}_{\mathbf{W}\mathrm{dir}2,2} \\ + \mathbf{U}_3 \mathbf{H}_{\mathbf{W}\mathrm{dir}3,1} + \mathbf{U}_4 \mathbf{H}_{\mathbf{W}\mathrm{dir}3,2} \end{bmatrix}, \qquad (119)$$

where the $\mathbf{w}$ subscript indicates that the vector is scaling/wavelet magnitudes. The updated $\mathbf{D}$ vector can be reconstructed to give field values at specific points by multiplication with $\mathbf{R}$,

$$\mathbf{D}_{\mathrm{dir}1} = \mathbf{R}\mathbf{D}_{\mathbf{W}\mathrm{dir}1} = \mathbf{R}\mathbf{D}_{\mathbf{W}\mathrm{dir}1} + \frac{\Delta t}{\Delta x \Delta y \Delta z} \begin{bmatrix} \mathbf{R}\mathbf{U}_1 \mathbf{H}_{\mathbf{W}\mathrm{dir}2,1} + \mathbf{R}\mathbf{U}_2 \mathbf{H}_{\mathbf{W}\mathrm{dir}2,2} \\ + \mathbf{R}\mathbf{U}_3 \mathbf{H}_{\mathbf{W}\mathrm{dir}3,1} + \mathbf{R}\mathbf{U}_4 \mathbf{H}_{\mathbf{W}\mathrm{dir}3,2} \end{bmatrix}. \qquad (120)$$

The $\mathbf{D}_{\mathrm{dir}1} = \mathbf{R}\mathbf{D}_{\mathbf{W}\mathrm{dir}1}$ term represents the field values at the equivalent grid points, not the scaling/wavelet coefficients. Thus, (120) is an equation that gives updated $\mathbf{D}$ field values from $\mathbf{H}$ scaling/wavelet coefficients. If the PEC condition is applied,

$$\mathbf{D}_{\mathrm{dir}1,\mathrm{PEC}} = \mathbf{I}_{\mathbf{P}}\mathbf{R}\mathbf{D}_{\mathbf{W}\mathrm{dir}1} = \mathbf{I}_{\mathbf{P}}\mathbf{R}\mathbf{D}_{\mathbf{W}\mathrm{dir}1}$$
$$+ \frac{\Delta t}{\Delta x \Delta y \Delta z} \begin{bmatrix} \mathbf{I}_{\mathbf{P}}\mathbf{R}\mathbf{U}_1 \mathbf{H}_{\mathbf{W}\mathrm{dir}2,1} + \mathbf{I}_{\mathbf{P}}\mathbf{R}\mathbf{U}_2 \mathbf{H}_{\mathbf{W}\mathrm{dir}2,2} \\ + \mathbf{I}_{\mathbf{P}}\mathbf{R}\mathbf{U}_3 \mathbf{H}_{\mathbf{W}\mathrm{dir}3,1} + \mathbf{I}_{\mathbf{P}}\mathbf{R}\mathbf{U}_4 \mathbf{H}_{\mathbf{W}\mathrm{dir}3,2} \end{bmatrix}. \qquad (121)$$

Transforming back into the wavelet domain,

$$\mathbf{D}_{\mathbf{W}\mathrm{dir}1,\mathrm{PEC}} = \mathbf{R}^{-1}\mathbf{I}_{\mathbf{P}}\mathbf{R}\mathbf{D}_{\mathbf{W}\mathrm{dir}1} = \mathbf{R}^{-1}\mathbf{I}_{\mathbf{P}}\mathbf{R}\mathbf{D}_{\mathbf{W}\mathrm{dir}1}$$
$$+ \frac{\Delta t}{\Delta x \Delta y \Delta z} \begin{bmatrix} \mathbf{R}^{-1}\mathbf{I}_{\mathbf{P}}\mathbf{R}\mathbf{U}_1 \mathbf{H}_{\mathbf{W}\mathrm{dir}2,1} + \mathbf{R}^{-1}\mathbf{I}_{\mathbf{P}}\mathbf{R}\mathbf{U}_2 \mathbf{H}_{\mathbf{W}\mathrm{dir}2,2} \\ + \mathbf{R}^{-1}\mathbf{I}_{\mathbf{P}}\mathbf{R}\mathbf{U}_3 \mathbf{H}_{\mathbf{W}\mathrm{dir}3,1} + \mathbf{R}^{-1}\mathbf{I}_{\mathbf{P}}\mathbf{R}\mathbf{U}_4 \mathbf{H}_{\mathbf{W}\mathrm{dir}3,2} \end{bmatrix}. \qquad (122)$$

If a new matrix $\mathbf{U}_{\mathrm{PEC}}$ is defined

$$\mathbf{U}_{\mathrm{PEC}} = \mathbf{R}^{-1}\mathbf{I}_{\mathbf{P}}\mathbf{R}\mathbf{U}, \qquad (123)$$

then a new update equation can be defined

$$\mathbf{D}_{\mathbf{W}\text{dir1,PEC}} = \mathbf{D}_{\mathbf{W}\text{dir,PEC1}} + \frac{\Delta t}{\Delta x \Delta y \Delta z} \begin{bmatrix} \mathbf{U}_{\mathbf{PEC1}}\mathbf{H}_{\mathbf{W}\text{dir2,1}} + \mathbf{U}_{\mathbf{PEC2}}\mathbf{H}_{\mathbf{W}\text{dir2,2}} \\ + \mathbf{U}_{\mathbf{PEC3}}\mathbf{H}_{\mathbf{W}\text{dir3,1}} + \mathbf{U}_{\mathbf{PEC4}}\mathbf{H}_{\mathbf{W}\text{dir3,2}} \end{bmatrix}.$$

(124)

The $\mathbf{R}^{-1}\mathbf{I}_{\mathbf{P}}\mathbf{R}\mathbf{D}_{\mathbf{W}\text{dir1}}$ condition does not have to be explicitly enforced because $\mathbf{D}_{\mathbf{W}\text{dir1}}$ represents the scaling/wavelet coefficients from the previous time step, where the condition has already been applied.

This technique provides a method that can be used to automatically apply the PEC boundary condition within the Haar-MRTD update equations. In this method, the $\mathbf{U}_{\mathbf{PEC}}$ matrices can be calculated before the simulation begins, and therefore do not add significant overhead to the field updates. In fact, this method can be expanded to allow other subcell effects to be automatically applied in the MRTD update scheme.

3.4   GENERAL SUBCELL EFFECTS IN HAAR-MRTD: COMPOSITE CELLS

In the previous section, the reconstruction/wavelet transform matrices were used to apply the PEC boundary condition directly in the MRTD update. If the reconstruction/deconstruction is applied to all of the fields in the update equation,

$$\begin{aligned} \mathbf{D}_{\text{dir1}} &= \mathbf{R}\mathbf{D}_{\mathbf{W}\text{dir1}} = \mathbf{R}\mathbf{D}_{\mathbf{W}\text{dir1}} \\ &+ \frac{\Delta t}{\Delta x \Delta y \Delta z} \begin{bmatrix} \mathbf{R}\mathbf{U}_1\mathbf{R}^{-1}\mathbf{R}\mathbf{H}_{\mathbf{W}\text{dir2,1}} + \mathbf{R}\mathbf{U}_2\mathbf{R}^{-1}\mathbf{R}\mathbf{H}_{\mathbf{W}\text{dir2,2}} \\ +\mathbf{R}\mathbf{U}_3\mathbf{R}^{-1}\mathbf{R}\mathbf{H}_{\mathbf{W}\text{dir3,1}} + \mathbf{R}\mathbf{U}_4\mathbf{R}^{-1}\mathbf{R}\mathbf{H}_{\mathbf{W}\text{dir3,2}} \end{bmatrix}, \end{aligned}$$

(125)

and the substitutions

$$\mathbf{H} = \mathbf{R}\mathbf{H}_{\mathbf{W}},$$

(126)

$$\mathbf{U}_{\mathbf{L}} = \mathbf{R}\mathbf{U}\mathbf{R}^{-1},$$

(127)

where the subscript **L** is used to refer to the update equation for local, pointwise, fields, is made, then

$$\mathbf{D}_{\text{dir1}} = \mathbf{D}_{\text{dir1}} + \frac{\Delta t}{\Delta x \Delta y \Delta z}\left[\begin{array}{c}\mathbf{U}_{\text{L1}}\mathbf{H}_{\text{dir2,1}} + \mathbf{U}_{\text{L2}}\mathbf{H}_{\text{dir2,2}} \\ + \mathbf{U}_{\text{L3}}\mathbf{H}_{\text{dir3,1}} + \mathbf{U}_{\text{L4}}\mathbf{H}_{\text{dir3,2}}\end{array}\right]. \qquad (128)$$

This equation is an update for the field values at the equivalent grid points from field values at equivalent grid points. It has been repeatedly stated that MRTD and FDTD are equivalent schemes, and (128) is the conversion from MRTD to FDTD. It should be noted, however, that the MRTD scheme can still be used to vary the wavelet resolution, and thus the number of equivalent grid points, on a cell-by-cell basis. This is an inherent property of MRTD that is not available in FDTD. The pointwise field update representation (128) provides the ability to manipulate the fields at individual equivalent grid points and then transform the fields back to the scaling/wavelet domain.

One application of this technique, and indeed the motivation for this technique, is the application of the PEC boundary condition at individual equivalent grid points that has already been presented. Another simple application of this technique is the addition of a current source at equivalent grid points. This method is equivalent to wavelet transforming a spatial source condition.

In the pointwise update scheme, a source, **J**, can be added at each equivalent grid point:

$$\mathbf{D}_{\text{dir1}} = \mathbf{D}_{\text{dir1}} + \frac{\Delta t}{\Delta x \Delta y \Delta z}\left[\begin{array}{c}\mathbf{U}_{\text{L1}}\mathbf{H}_{dir2,1} + \mathbf{U}_{\text{L2}}\mathbf{H}_{\text{dir2,2}} \\ + \mathbf{U}_{\text{L3}}\mathbf{H}_{\text{dir3,1}} + \mathbf{U}_{\text{L4}}\mathbf{H}_{\text{dir3,2}}\end{array}\right] + \mathbf{J}. \qquad (129)$$

Multiplying (129) by $\mathbf{R}^{-1}$ converts the equation back to the wavelet domain, yielding

$$\mathbf{D}_{\text{Wdir1}} = \mathbf{R}^{-1}\mathbf{D}_{\text{dir1}} + \frac{\Delta t}{\Delta x \Delta y \Delta z}\mathbf{R}^{-1}\left[\begin{array}{c}\mathbf{U}_{\text{L1}}\mathbf{H}_{\text{dir2,1}} + \mathbf{U}_{\text{L2}}\mathbf{H}_{\text{dir2,2}} \\ + \mathbf{U}_{\text{L3}}\mathbf{H}_{\text{dir3,1}} + \mathbf{U}_{\text{L4}}\mathbf{H}_{\text{dir3,2}}\end{array}\right] + \mathbf{R}^{-1}\mathbf{J}$$

$$= \mathbf{D}_{\text{Wdir1}} + \frac{\Delta t}{\Delta x \Delta y \Delta z}\left[\begin{array}{c}\mathbf{U}_1\mathbf{H}_{\text{Wdir2,1}} + \mathbf{U}_2\mathbf{H}_{\text{Wdir2,2}} \\ + \mathbf{U}_3\mathbf{H}_{\text{Wdir3,1}} + \mathbf{U}_4\mathbf{H}_{\text{Wdir3,2}}\end{array}\right] + \mathbf{R}^{-1}\mathbf{J}. \qquad (130)$$

This equation can be used to apply an arbitrarily placed source into the MRTD cell.

One of the advantages of the Yee-FDTD scheme is that its popularity and longevity has led to the development of a large number of techniques that can extend its use. Some examples of techniques that have been developed for Yee-FDTD are the modeling of thin wires [21], narrow slots [22], curved structures (with a locally conformal grid) [23], thin material sheets [24], dispersive surfaces (such as thin metals) [25], SPICE circuits [26], local field correction [27], and lumped elements [28]. The result of all of these techniques is modified FDTD update equations for the areas where the effect is applied. The surrounding fields are updated normally. Using the technique that has been presented here, it is possible to directly bridge the modified FDTD update equations to the MRTD technique.

An example of how this technique can be applied to bridge the pointwise FDTD modifications to the MRTD technique is presented here with lumped elements. The procedure for representing lumped elements in FDTD is presented in [28], and a brief overview is given here.

A lumped element, here specified to be a resistor, capacitor, or inductor, can be represented in Ampere's law as a current source $\mathbf{J_L}$,

$$\nabla \times \mathbf{H}(t) = \frac{\partial \mathbf{D}(t)}{\partial t} + \mathbf{J_L}(t). \tag{131}$$

In this case currents due to loss and impressed currents are neglected for simplicity. For each of the lumped elements listed above, a relationship can be determined for the current as a function of voltage. The $\mathbf{D_z}$ component update will be shown here, although this procedure can be easily modified for any direction. Including the lumped element current, the z component of (131) is

$$\frac{\partial D_z}{\partial t} = \frac{\partial H_y}{\partial x} - \frac{\partial H_x}{\partial y} - \frac{I_z}{\Delta x \Delta y}, \tag{132}$$

noting that

$$J_{L,z} = \frac{I_z}{\Delta x \Delta y}. \tag{133}$$

I_z can be determined for each element. For resistors,

$$I_{z,R} = \frac{V}{R} = \frac{1}{R} \int_{\Delta z} E \partial z; \qquad (134)$$

for capacitors,

$$I_{z,C} = C \frac{\partial V}{\partial t} = C \frac{\partial \int_{\Delta z} E \partial z}{\partial t}; \qquad (135)$$

and for inductors,

$$I_{z,C} = \frac{1}{I} \int V \partial t = \frac{1}{I} \int \int_{\Delta z} E \partial z \partial t. \qquad (136)$$

For any of these cases, the MRTD update equations can be determined by inserting the current relationship into Ampere's law and applying the method of moments. However, this method is somewhat difficult. An example is presented for the resistor.

The voltage across a single equivalent grid point can be simply calculated as

$$V_{i,j,k} = E_{i,j,k} \Delta z. \qquad (137)$$

However, for Ampere's law, the update is performed at the $n + 1/2$ time step, while the E field is known only for the n and $n + 1$ time steps. Similar to the method that is used for Ohmic losses, the semiimplicit approximation can be applied. In this case,

$$V_{i,j,k} = \frac{n+1 E_{i,j,k} + n E_{i,j,k}}{2} \Delta z. \qquad (138)$$

When the method of moments is applied to (132), with the voltage–current relationship (138), the resulting update equation is

$$\begin{aligned}
n+1 \mathbf{D}_{z,i,j,k} = {}& n \mathbf{D}_{z,i,j,k} + \mathbf{L} \left(n+1 \mathbf{E}_{z,i,j,k} + n \mathbf{E}_{z,i,j,k} \right) \\
& + \frac{\Delta t}{\Delta x \Delta y \Delta z} \left[\begin{array}{l} \mathbf{U}^{D_z}_{H_x,1 n+1/2} \mathbf{H}_{x,i,j+1,k} + \mathbf{U}^{D_z}_{H_x,0 n+1/2} \mathbf{H}_{x,i,j,k} \\ + \mathbf{U}^{D_z}_{H_y,1 n+1/2} \mathbf{H}_{y,i+1,j,k} + \mathbf{U}^{D_z}_{H_y,0 n+1/2} \mathbf{H}_{y,i,j,k} \end{array} \right]. \quad (139)
\end{aligned}$$

The entries of $\mathbf{L}$ have the form,

$$L_{ij} = \int_{\mathbf{r}} L(\mathbf{r}) \Gamma_i \Gamma_j \partial \mathbf{r}, \tag{140}$$

with

$$L(\mathbf{r}) = \begin{cases} \dfrac{\Delta z}{2R\Delta x \Delta y} & \text{resistor location} \\ 0 & \text{elsewhere} \end{cases}, \tag{141}$$

if Δx, Δy, and Δz are the size of the equivalent grid points. Two difficulties are immediately apparent with the equation. First, it includes both $\mathbf{D}$ and $\mathbf{E}$ coefficients, where before $\mathbf{E}$ coefficients were updated directly from the $\mathbf{D}$ fields. Secondly, $\mathbf{D}$ fields are updated using $\mathbf{E}$ fields at the same time step.

The first of these difficulties can be easily accounted for using the constitutive relationship. In (57), the electric field is determined from the D field. In the Haar case, this relationship becomes

$$_n\mathbf{E}_{d,i,j,k} = \mathbf{U}^{E_d}_{D_d,i,j,k,a,b,c}{}_n\mathbf{D}_{d,i,j,k} = \varepsilon_{d,i,j,k}{}_n\mathbf{D}_{d,i,j,k}. \tag{142}$$

Because no neighboring field values are required to determine the $\mathbf{E}$ coefficients from the $\mathbf{D}$ coefficients, the transformation can be applied directly to the update equation. In this case,

$$
\begin{aligned}
_{n+1}&\mathbf{E}_{z,i,j,k} \\
&= \varepsilon_{d,i,j,k}{}_{n+1}\mathbf{D}_{z,i,j,k} \\
&= \varepsilon_{d,i,j,k}{}_{n}\mathbf{D}_{z,i,j,k} \\
&\quad + \frac{\Delta t}{\Delta x \Delta y \Delta z}\varepsilon_{d,i,j,k}
\begin{bmatrix}
\mathbf{U}_{H_x,1\,n+1/2}\mathbf{H}_{x,i,j+1,k} + \mathbf{U}_{H_x,0\,n+1/2}\mathbf{H}_{x,i,j,k} \\
+\, \mathbf{U}_{H_y,1\,n+1/2}\mathbf{H}_{y,i+1,j,k} + \mathbf{U}_{H_y,0\,n+1/2}\mathbf{H}_{y,i,j,k}
\end{bmatrix} \\
&= \mathbf{E}_{z,i,j,k} + \frac{\Delta t}{\Delta x \Delta y \Delta z}\varepsilon_{d,i,j,k}
\begin{bmatrix}
\mathbf{U}_{H_x,1\,n+1/2}\mathbf{H}_{x,i,j+1,k} + \mathbf{U}_{H_x,0\,n+1/2}\mathbf{H}_{x,i,j,k} \\
+\, \mathbf{U}_{H_y,1\,n+1/2}\mathbf{H}_{y,i+1,j,k} + \mathbf{U}_{H_y,0\,n+1/2}\mathbf{H}_{y,i,j,k}
\end{bmatrix}.
\end{aligned}
\tag{143}
$$

Using this update, (139) becomes

$$
\begin{aligned}
_{n+1}\mathbf{E}_{z,i,j,k} = {}&_{n}\mathbf{E}_{z,i,j,k} + \mathbf{L}\left(_{n+1}\mathbf{E}_{z,i,j,k} + {}_{n}\mathbf{E}_{z,i,j,k}\right) \\
&+ \frac{\Delta t}{\Delta x \Delta y \Delta z}\varepsilon
\begin{bmatrix}
\mathbf{U}_{H_x,1\,n+1/2}\mathbf{H}_{x,i,j+1,k} + \mathbf{U}_{H_x,0\,n+1/2}\mathbf{H}_{x,i,j,k} \\
+\mathbf{U}_{H_y,1\,n+1/2}\mathbf{H}_{y,i+1,j,k} + \mathbf{U}_{H_y,0\,n+1/2}\mathbf{H}_{y,i,j,k}
\end{bmatrix}.
\end{aligned}
\tag{144}
$$

Collecting terms

$$
\begin{aligned}
(\mathbf{I} - \mathbf{L})_{n+1}\mathbf{E}_{z,i,j,k} = {}&(\mathbf{I} + \mathbf{L})_{n}\mathbf{E}_{z,i,j,k} + \frac{\Delta t}{\Delta x \Delta y \Delta z}\varepsilon \\
&\times
\begin{bmatrix}
\mathbf{U}_{H_x,1\,n+1/2}\mathbf{H}_{x,i,j+1,k} + \mathbf{U}_{H_x,0\,n+1/2}\mathbf{H}_{x,i,j,k} \\
+\mathbf{U}_{H_y,1\,n+1/2}\mathbf{H}_{y,i+1,j,k} + \mathbf{U}_{H_y,0\,n+1/2}\mathbf{H}_{y,i,j,k}
\end{bmatrix},
\end{aligned}
\tag{145}
$$

and solving for $_{n+1}\mathbf{E}_{x,i,j,k}$

$$
\begin{aligned}
_{n+1}\mathbf{E}_{z,i,j,k} = {}&(\mathbf{I} - \mathbf{L})^{-1}(\mathbf{I} + \mathbf{L})_{n}\mathbf{E}_{z,i,j,k} + \frac{\Delta t}{\Delta x \Delta y \Delta z}(\mathbf{I} - \mathbf{L})^{-1}\varepsilon \\
&\times
\begin{bmatrix}
\mathbf{U}_{H_x,1\,n+1/2}\mathbf{H}_{x,i,j+1,k} + \mathbf{U}_{H_x,0\,n+1/2}\mathbf{H}_{x,i,j,k} \\
+\mathbf{U}_{H_y,1\,n+1/2}\mathbf{H}_{y,i+1,j,k} + \mathbf{U}_{H_y,0\,n+1/2}\mathbf{H}_{y,i,j,k}
\end{bmatrix}.
\end{aligned}
\tag{146}
$$

A similar procedure can be used to determine the updates for capacitors and inductors. This procedure is not difficult and most of the operations involved are matrix multiplications that can be performed before the start of simulation. The only integration required is to determine the $\mathbf{L}$ matrix. However, there is another method that can be used in this case that removes the need to calculate the $\mathbf{L}$ matrix.

The E_z update equation in FDTD for a resistor is [28]

$$
_{n+1}E_{z,i,j,k} = \left(\frac{1 - \frac{\Delta t \Delta z}{2R\varepsilon \Delta x \Delta y}}{1 + \frac{\Delta t \Delta z}{2R\varepsilon \Delta x \Delta y}}\right)_{n}E_{z,i,j,k}
\tag{147}
$$

$$
+ \left(\frac{\frac{\Delta t}{\varepsilon}}{1 + \frac{\Delta t \Delta z}{2R\varepsilon \Delta x \Delta y}}\right)
\begin{bmatrix}
\frac{_{n+1/2}H_{y,i+1,j,k} - {}_{n+1/2}H_{y,i,j,k}}{\Delta x} \\
- \frac{_{n+1/2}H_{x,i,j+1,k} - {}_{n+1/2}H_{x,i,j,k}}{\Delta y}
\end{bmatrix}.
\tag{148}
$$

Using (128), two additional lumped element matrices can be created that apply the lumped element formulation in (147). The two lumped element matrices $\mathbf{L_1}$ and

L_2 are diagonal with

$$L_{1,i,i} = \begin{cases} \varepsilon \left(\dfrac{1 - \frac{\Delta t \Delta z}{2 R \varepsilon \Delta x \Delta y}}{1 + \frac{\Delta t \Delta z}{2 R \varepsilon \Delta x \Delta y}} \right) & \text{resistor equivalent grid points} \\ 1 & \text{elsewhere} \end{cases}, \qquad (149)$$

$$L_{2,i,i} = \begin{cases} \left(\dfrac{1}{1 + \frac{\Delta t \Delta z}{2 R \varepsilon \Delta x \Delta y}} \right) & \text{resistor equivalent grid points} \\ 1 & \text{elsewhere} \end{cases}. \qquad (150)$$

L_1 and L_2 can be inserted into (128)

$$\mathbf{D}_{\text{dir1}} = \mathbf{L_1} \mathbf{D}_{\text{dir1}} + \frac{\Delta t}{\Delta x \Delta y \Delta z} \mathbf{L_2} \begin{bmatrix} \mathbf{U_{L1}} \mathbf{H}_{\text{dir2,1}} + \mathbf{U_{L2}} \mathbf{H}_{\text{dir2,2}} \\ + \mathbf{U_{L3}} \mathbf{H}_{\text{dir3,1}} + \mathbf{U_{L4}} \mathbf{H}_{\text{dir3,2}} \end{bmatrix}, \qquad (151)$$

where again, for simplicity, the subscripts have been dropped. By rotating the values of Δx, Δy, and Δz, $\mathbf{L}$ can be used for a resistor in any direction. Converting back to the wavelet domain,

$$\begin{aligned} \mathbf{D}_{\text{Wdir1}} = \mathbf{R}^{-1} \mathbf{L_1} \mathbf{R}^{-1} \mathbf{D}_{\text{Wdir1}} + \frac{\Delta t}{\Delta x \Delta y \Delta z} \mathbf{R}^{-1} \mathbf{L_2} \\ \times \begin{bmatrix} \mathbf{U_{L1}} \mathbf{R} \mathbf{H}_{\text{Wdir2,1}} + \mathbf{U_{L2}} \mathbf{R} \mathbf{H}_{\text{Wdir2,2}} \\ + \mathbf{U_{L3}} \mathbf{R} \mathbf{H}_{\text{Wdir3,1}} + \mathbf{U_{L4}} \mathbf{R} \mathbf{H}_{\text{Wdir3,2}} \end{bmatrix}. \end{aligned} \qquad (152)$$

These matrices can be combined so that each field is only multiplied by one matrix for each update

$$\mathbf{D}_{\text{Wdir1}} = \mathbf{U_R} \mathbf{D}_{\text{Wdir1}} + \frac{\Delta t}{\Delta x \Delta y \Delta z} \begin{bmatrix} \mathbf{U_{R1}} \mathbf{H}_{\text{Wdir2,1}} + \mathbf{U_{R2}} \mathbf{H}_{\text{Wdir2,2}} \\ + \mathbf{U_{R3}} \mathbf{H}_{\text{Wdir3,1}} + \mathbf{U_{R4}} \mathbf{H}_{\text{Wdir3,2}} \end{bmatrix}. \qquad (153)$$

Using this method, resistors can be inserted into the MRTD grid at any equivalent grid point, and modified update equations can be found before simulation through simple matrix multiplications. One other important note is that this formulation introduced a new update matrix for the current $\mathbf{D}$ field. This does add one extra matrix multiplication to the field update, but, as is shown in the next chapter, the general MRTD update including UPML includes this matrix.

Similar updates can be derived for capacitors and inductors. Capacitors are modeled in FDTD using

$$_{n+1}E_{z,i,j,k} = {}_nE_{z,i,j,k} + \left(\frac{\frac{\Delta t}{\varepsilon}}{1 + \frac{C\Delta z}{\varepsilon \Delta x \Delta y}}\right)\left[\begin{array}{c} \frac{_{n+1/2}H_{y,i+1,j,k} - {}_{n+1/2}H_{y,i,j,k}}{\Delta x} \\ -\frac{_{n+1/2}H_{x,i,j+1,k} - {}_{n+1/2}H_{x,i,j,k}}{\Delta y} \end{array}\right],$$

(154)

so, in this case, only one $\mathbf{L}$ matrix needs to be determined,

$$L_{2,i,i} = \begin{cases} \left(\dfrac{1}{1 + \frac{C\Delta z}{\varepsilon \Delta x \Delta y}}\right) & \text{capacitor equivalent grid points} \\ 1 & \text{elsewhere} \end{cases}.$$

(155)

In FDTD, inductors are represented using,

$$_{n+1}E_{z,i,j,k} = {}_nE_{z,i,j,k} + \frac{\Delta t}{\varepsilon}\left[\begin{array}{c} \frac{_{n+1/2}H_{y,i+1,j,k} - {}_{n+1/2}H_{y,i,j,k}}{\Delta x} \\ -\frac{_{n+1/2}H_{x,i,j+1,k} - {}_{n+1/2}H_{x,i,j,k}}{\Delta y} \end{array}\right]$$
$$-\frac{\Delta z(\Delta t)^2}{\varepsilon L \Delta x \Delta y}\sum_{m=1}^{n} {}_mE_{z,i,j,k},$$

(156)

where it is noted that the sum can be represented as a single value that is augmented at each time step. In this case, both $\mathbf{L_1}$ and $\mathbf{L_2}$ are the identity matrix, but another matrix must be created. In this case, the sum term is similar to the previously presented current source. If (128) is modified

$$\mathbf{D}_{\text{dir1}} = \mathbf{D}_{\text{dir1}} + \frac{\Delta t}{\Delta x \Delta y \Delta z}\left[\begin{array}{c} \mathbf{U_{L1}H}_{\text{dir2,1}} + \mathbf{U_{L2}H}_{\text{dir2,2}} \\ + \mathbf{U_{L3}H}_{\text{dir3,1}} + \mathbf{U_{L4}H}_{\text{dir3,2}} \end{array}\right] + \mathbf{L}\sum_{m=0}^{n}\mathbf{D},$$

(157)

with

$$L_{i,i} = \begin{cases} \dfrac{\Delta z(\Delta t)^2}{\varepsilon L \Delta x \Delta y} & \text{inductor equivalent grid points} \\ 0 & \text{elsewhere} \end{cases},$$

(158)

then, using the same procedure as above, (157) can be converted into the wavelet domain. In the case of the inductor, an additional vector, $\sum_{m=0}^{n} \mathbf{D_W}$ must be used to calculate the updates. It is noted that it is only necessary to store this vector in cells that contain inductors.

CHAPTER 4

Other Techniques Necessary for Simulation: UPML, Variable Gridding, Source Excitation and Time/Space Adaptive Gridding

In the previous chapter, a method is presented that allows a variety of elements that are smaller than a single MRTD cell to be simulated using the Haar MRTD method. Unlike the method presented in [22], this technique treats the entire space as an MRTD grid, using scaling and wavelet functions exclusively to represent the field. The method was created to model subcell PEC structures, as most microwave structures are simulated using only PEC and dielectric media. In addition, it is shown that the method can be generalized to allow the modeling of several other subcell effects; a specific example is provided for lumped elements. The MRTD method that includes subcell elements and varying dielectrics within individual

MRTD cells is termed the composite-cell method. This is because, instead of homogenous cells that are used in most MRTD simulations, these cells can include complex structures. The advantage of this method is that these subcell structures can be treated within the MRTD framework, allowing the adaptive MRTD grid to be used to effectively model these structures.

In order to perform simulations using the MRTD method, there are a few more topics that must be addressed. The first is grid truncation. While a variety of absorbing boundary conditions can be implemented, the treatment in this book focuses exclusively on the UPML. It is one of the best absorbers (if computationally intensive), and can be easily implemented using the MRTD framework previously presented. In addition, grid excitation is necessary to inject signals into the computational space. The treatment in this chapter follows on the impressed current source formulation presented in the previous chapter. Finally, a method of varying grid size is presented, which allows the modeling of complex structures more accurately and efficiently. Like in the previous chapter, this section discusses many techniques that were originally developed for the FDTD method.

4.1   ARBITRARY WAVELET RESOLUTION UPML

In order to perform simulations of microwave structures, a method of truncating the computational space with a reflectionless (or low reflection) condition is necessary. These absorbing boundary conditions allow the computational space to closely surround the simulated structure. Without such a condition, signals that reach the outer boundary (represented as a PEC or PMC) will reflect into the space. This renders the field values observed in the simulated structure useless; there is no way to differentiate between the reflected and incident signal. The most widely used method of truncating the computational space is the perfectly matched layer, which has the properties of extremely high loss while matching arbitrary incident signals.

The UPML [19] was first developed as an alternative to the Berenger PML [18] that did not require nonphysical field splitting. The UPML is expressed as a material with a carefully designed permittivity and permeability tensor [14]

$$\bar{\bar{\varepsilon}}\,(\omega) = \bar{\bar{\mu}}\,(\omega) = \begin{bmatrix} \dfrac{s_y s_z}{s_x} & 0 & 0 \\ 0 & \dfrac{s_x s_z}{s_y} & 0 \\ 0 & 0 & \dfrac{s_x s_y}{s_z} \end{bmatrix}, \tag{158}$$

where

$$s_n = k_n + \frac{\sigma_n}{j\omega\varepsilon} \qquad n = x,\, y,\, z. \tag{159}$$

The parameter σ_n represents the loss inside the UPML and k_n is a matching parameter that can be used to fine tune the UPML performance.

The permittivity and permeability tensors inside the UPML are frequency dependent, and are therefore difficult to represent in a time domain scheme. If the constitutive relationship

$$D_x\,(\omega) = \varepsilon \frac{s_z}{s_x} E_x\,(\omega) \qquad D_y\,(\omega) = \varepsilon \frac{s_x}{s_y} E_y\,(\omega) \qquad D_z\,(\omega) = \varepsilon \frac{s_y}{s_z} E_z\,(\omega) \tag{160}$$

is defined, then a coupled set of time domain equations can be defined to update the electric and magnetic fields. The derivation for the electric fields is shown here, the magnetic field updates are found in the same manner.

The time domain differential equation for the D field is

$$\begin{bmatrix} \dfrac{\partial H_z}{\partial y} - \dfrac{\partial H_y}{\partial z} \\ \dfrac{\partial H_x}{\partial z} - \dfrac{\partial H_z}{\partial x} \\ \dfrac{\partial H_y}{\partial x} - \dfrac{\partial H_x}{\partial y} \end{bmatrix} = \begin{bmatrix} k_y & 0 & 0 \\ 0 & k_z & 0 \\ 0 & 0 & k_z \end{bmatrix} \frac{\partial}{\partial t} \begin{bmatrix} D_x \\ D_y \\ D_z \end{bmatrix} + \frac{1}{\varepsilon} \begin{bmatrix} \sigma_y & 0 & 0 \\ 0 & \sigma_z & 0 \\ 0 & 0 & \sigma_z \end{bmatrix} \begin{bmatrix} D_x \\ D_y \\ D_z \end{bmatrix}. \tag{161}$$

This system is very similar to Ampere's law when Ohmic loss is included (the only difference is the constant that precedes the time derivative). It can be discretized using the procedure outlined in Chapter 2. For example, the D_x component of

(161) is

$$\frac{\partial H_z}{\partial y} - \frac{\partial H_y}{\partial z} = k_y \frac{\partial D_x}{\partial t} + \frac{\sigma_y}{\varepsilon} D_x, \tag{162}$$

and the time localized form is

$$_{n+1}D_x = {}_nD_x - \frac{\Delta t \sigma_y}{k_y \varepsilon} \left(\frac{{}_{n+1}D_x + {}_nD_x}{2} \right) + \frac{\Delta t}{k_y} \left(\frac{\partial_{n+1/2}H_z}{\partial y} - \frac{\partial_{n+1/2}H_y}{\partial z} \right). \tag{163}$$

Combining D terms,

$$_{n+1}D_x = {}_nD_x \left(\frac{2k_y\varepsilon - \Delta t\sigma_y}{2k_y\varepsilon + \Delta t\sigma_y} \right) + \left(\frac{2k_y\varepsilon\Delta t}{2k_y\varepsilon + \Delta t\sigma_y} \right) \left(\frac{\partial_{n+1/2}H_z}{\partial y} - \frac{\partial_{n+1/2}H_y}{\partial z} \right). \tag{164}$$

Next, space localization can be performed, and the update matrices can be computed. The $\mathbf{D_x}$ update is

$$_{n+1}\mathbf{D}_{x,i,j,k} = \mathbf{U}_{D_x,}^{D_x,}{}_n\mathbf{D}_{x,i,j,k} + \frac{1}{\Delta x \Delta y \Delta z}$$
$$\times \left[\sum_l \sum_m \mathbf{U}_{H_y,m}^{D_x}{}_{n+1/2}\mathbf{H}_{y,i,j+l,k+m} + \sum_m \mathbf{U}_{H_z,m}^{D_x}{}_{n+1/2}\mathbf{H}_{z,i,j+m,k} \right], \tag{165}$$

where

$$U_{D_x,i,j}^{D_x} = \left\langle {}_{D_x}\Gamma_i, \left(\frac{2k_y\varepsilon - \Delta t\sigma_y}{2k_y\varepsilon + \Delta t\sigma_y} \right)_{D_x} \Gamma_j \right\rangle, \tag{166}$$

$$U_{H_y,m,i,j}^{D_x} = \left\langle {}_{D_x}\Gamma_i, -\left(\frac{2k_y\varepsilon\Delta t}{2k_y\varepsilon + \Delta t\sigma_y} \right) \frac{\partial}{\partial z} {}_{H_y}\Gamma_j\big|_m \right\rangle, \tag{167}$$

$$U_{H_z,m,i,j}^{D_x} = \left\langle {}_{D_x}\Gamma_i, \left(\frac{2k_y\varepsilon\Delta t}{2k_y\varepsilon + \Delta t\sigma_y} \right) \frac{\partial}{\partial y} {}_{H_z}\Gamma_j\big|_m \right\rangle. \tag{168}$$

Similar equations can be found for the $\mathbf{D_y}$ and $\mathbf{D_z}$ components.

The relationship between the E and D fields inside the PML is also more complicated than in the isotropic nondispersive case that has already been presented.

When the D_x relationship of (160) is converted to the time domain [19],

$$\frac{\partial}{\partial t}(k_x D_x) + \frac{\sigma_x}{\varepsilon} D_x = \varepsilon \left[\frac{\partial}{\partial t}(k_z E_x) + \frac{\sigma_z}{\varepsilon} E_x \right] \tag{169}$$

must be solved to determine the updated E fields from D fields.

The time localized form of (169) is

$$k_x \left({}_{n+1}D_x - {}_n D_x \right) + \frac{\Delta t \sigma_x}{\varepsilon} \frac{{}_{n+1}D_x + {}_n D_x}{2}$$
$$= \varepsilon \left[k_z \left({}_{n+1}E_x - {}_n E_x \right) + \frac{\Delta t \sigma_z}{\varepsilon} \frac{{}_{n+1}E_x + {}_n E_x}{2} \right]. \tag{170}$$

Collecting terms and solving for ${}_{n+1}E_x$,

$${}_{n+1}E_x = \left(\frac{2\varepsilon k_z - \sigma_z \Delta t}{2\varepsilon k_z + \sigma_z \Delta t} \right) {}_n E_x + \left[\frac{1}{(2\varepsilon k_z + \sigma_z \Delta t)\varepsilon} \right] \left[\begin{array}{l} (2\varepsilon k_x + \sigma_x \Delta t)\, {}_{n+1}D_x \\ - (2\varepsilon k_x - \sigma_x \Delta t)\, {}_n D_x \end{array} \right]. \tag{171}$$

Solving for the $\mathbf{E_x}$ coefficients,

$${}_{n+1}\mathbf{E}_{x,i,j,k} = \sum_m \mathbf{U}^{E_x}_{E_x,i,j,m}\, {}_n \mathbf{E}_{x,i,j,k+m} + \sum_a \sum_b \sum_c \mathbf{U}^{E_x}_{D_x,i,j,k,a,b,c}\, {}_n \mathbf{D}_{x,i+a,j+b,k+c}, \tag{172}$$

where

$$U^{E_x}_{E_x i,j,m} = \left\langle {}_{E_x}\Gamma_i, \left(\frac{2\varepsilon k_z - \sigma_z \Delta t}{2\varepsilon k_z + \sigma_z \Delta t} \right) {}_{E_x}\Gamma_j \big|_m \right\rangle, \tag{173}$$

$$U^{E_x}_{D_x+i,j} = \left\langle {}_{E_x}\Gamma_i, \frac{1}{\varepsilon} \left(\frac{2\varepsilon k_x + \sigma_x \Delta t}{2\varepsilon k_z + \sigma_z \Delta t} \right) {}_{E_x}\Gamma_j \big|_{a,b,c} \right\rangle, \tag{174}$$

$$U^{E_x}_{D_x i,j} = \left\langle {}_{E_x}\Gamma_i, -\frac{1}{\varepsilon} \left(\frac{2\varepsilon k_x - \sigma_x \Delta t}{2\varepsilon k_z + \sigma_z \Delta t} \right) {}_{E_x}\Gamma_j \big|_{a,b,c} \right\rangle, \tag{175}$$

and, again, similar equations can be found for the other directions. A process similar to the one presented here can be used to find the updates for the magnetic fields.

It has been left until this point to discuss the values of σ and k that should be used with this scheme. The technique for implementing the UPML that has been presented here is derived in an equivalent manner to the FDTD UPML [19] (until the last step, where the updates for the scaling/wavelet coefficients are found). The values of σ and k can be chosen in exactly the same manner as an FDTD implementation. In this case, $\sigma_a(n)$, where a is x, y, or z, is nonzero only within a predetermined distance d of the a normal outer boundaries. Usually, a width of 10 cells is sufficient for the UPML. Inside this boundary, $\sigma_a(n)$ is varied from zero to a maximum value $\sigma_{\max}$. A polynomial grading works well,

$$\sigma_a(n) = \left(n/d\right)^m \sigma_{q,\max}. \tag{176}$$

Similarly, k can be varied from 1 in the non-UPML region to $k_{\max}$ at the outer edge of the grid.

There is one important difference between the FDTD and MRTD implementations of the UPML. In FDTD, the σ values are discretized and applied only at FDTD grid points. In MRTD, however, the function $\sigma_a(n)$ is used when calculating the update equations, and thus the variation in $\sigma_a(n)$ across a cell is accounted for. This method could also be applied to FDTD, by determining FDTD coefficients as in S-MRTD.

It is important to note the similarity between the field updates for the PML and the field updates for fields outside the PML. If $\sigma = 0$ and $k = 1$, the equations are the same as for the PML free case. For implementation of the PML in a computer, it is convenient to use the PML formulation everywhere, and to use only nonzero σ values inside the PML region. As an added convenience, by using this formulation, the PML does not need to cover an entire cell. For example, in the $r_{\max} = 1$ case, there are four equivalent grid points per direction. If at least 10 equivalent cells of PML coverage are required, then this scheme permits exactly 10 to be used, instead of 12 (for three cells entirely covered with PML).

4.2   NONUNIFORM GRID IN MRTD

In the FDTD method, one common technique for conformal meshing is nonuniform gridding [29]. For a fixed grid size, it is easy to pick a structure (for example, parallel transmission lines) where a uniform grid (constant Δx, Δy, Δz) cannot align with the structure at all points. If, instead, the grid spacing can vary with position, many more structures can be accurately represented in the FDTD grid. Of course, curved structures must still be stair-stepped, but even with these structures, a more accurate grid using fewer grid points is possible. The FDTD algorithm depends on neighboring cells having identical dimensions at intersections, and if the grid is varied so that the grid size is a function of all three coordinate directions, the grid becomes incompatible with FDTD. Therefore, FDTD nonuniform grids must vary the spacing as a function of individual directions only. This means

$$\Delta x = \Delta x(i) \qquad \Delta y = \Delta y(j) \qquad \Delta z = \Delta z(k). \qquad (177)$$

One example of a grid that uses this scheme is presented in Fig. 14.

The 2-D grid that is presented shows the positions of the E_x, E_y, and H_z fields. The value of Δx or Δy must be computed based on the field update. For the space derivatives of fields, the value of Δx or Δy is the separation between the field components. Thus, for electric field space derivatives, the Δx or Δy value is simply the grid spacing. However, for magnetic field spatial derivatives, the value is instead the average of the Δ values for the cells containing each magnetic field. For example, if the H_z coefficient centered at Δx_3, Δy_3 in Fig. 14 is updated, the E_y field coefficients required are the ones centered at Δy_3 and located on either end of Δx_3. For this update, $\Delta y = \Delta y_3$. But when the E_y field component centered at Δy_3 and between Δx_3 and Δx_4 is updated, the H_z fields required are those centered at Δx_3, Δy_3 and Δx_4, Δy_3. In this case, $\Delta x = (\Delta x_3 + \Delta x_4)/2$.

It would be convenient to have a similar capability for MRTD. While the MRTD adaptive grid does allow the resolution to vary on a cell-by-cell basis, the equivalent grid size is always

$$\Delta n_{equiv} = \frac{\Delta n}{2^{1+r_n}}. \qquad (178)$$

Using this grid size, it is difficult to align the grid with realistic structures. Furthermore, increasing the resolution until an equivalent grid point is close to a feature

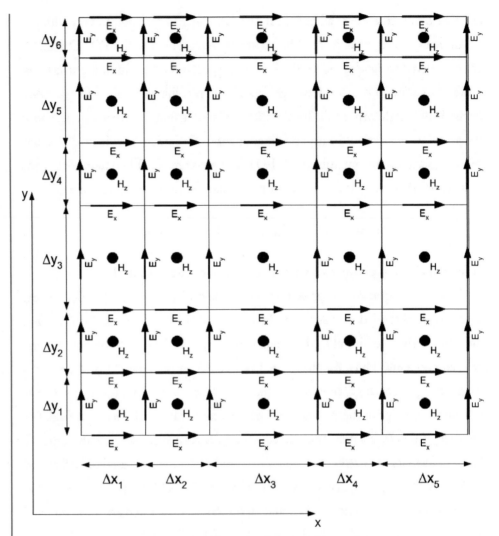

FIGURE 14: Two-dimensional FDTD nonuniform grid example.

being modeled is an inefficient use of the adaptive grid. A more effective grid would use both nonuniform and adaptive gridding.

The most difficult aspect of implementing a nonuniform grid in the MRTD method is determining the correct Δx, Δy, and Δz values to use for each field. In MRTD, the cell size does not appear explicitly in the field update (for the formulation presented here), but is an implicit part of determining the update coefficients. In the derivation presented in Chapter 2, a formula for the offset between the electric and magnetic fields is given as

$$s_d = \frac{\Delta_d}{2^{r_{d,\max}+2}}. \tag{179}$$

In this formula, $r_{d,\max}$ is the maximum resolution in the given direction. It was discussed that in the general 3-D scheme, the fields being updated (the time derivative terms in Faraday's and Ampere's laws) are offset from the other fields in the update (the spatial derivative terms) only in the direction of their spatial derivative. If this scheme is applied consistently for all electric and magnetic field coefficients, there is a specific arrangement of grid points that must be used. This arrangement is outlined here.

The convention chosen for this work is that all fields are offset relative to a grid that is defined by (177). The field components can be indexed to each of these grid points, but the actual domain of each function, defined by the Haar-MRTD scaling function (this discussion can be easily extended to general wavelet basis by defining the domain as the spacing between the centers of the scaling functions, centered on each scaling function), begins at the offset from the grid point. The grid is consistent with the conditions above if the electric fields are offset by s in their coordinate directions, and the magnetic fields are offset by s_d in the two directions normal to their coordinate direction. An example of a grid that meets these criteria is demonstrated in Fig. 15. It should be noted that while the offset can be made relative to another point (for example, the magnetic fields could be offset in their coordinate directions and the electric field in the normals) the resulting arrangement of the fields relative to each other must be maintained.

The offset of (177) denotes the start of the scaling function for the electric or magnetic field, but does not specify the size of the scaling function. If the size of each scaling function is chosen to simply as $\Delta x(i) \times \Delta y(j) \times \Delta z(k)$, then the basis functions for a single field can overlap. In Fig. 15, a 2-D view of an MRTD grid for $r_{\max} = 0$ (in both directions) is used to demonstrate variable gridding. For $r_{\max} = 0, s_x = s_y = 1/4$. In the figure, the positions of the E_z and H_z equivalent grid

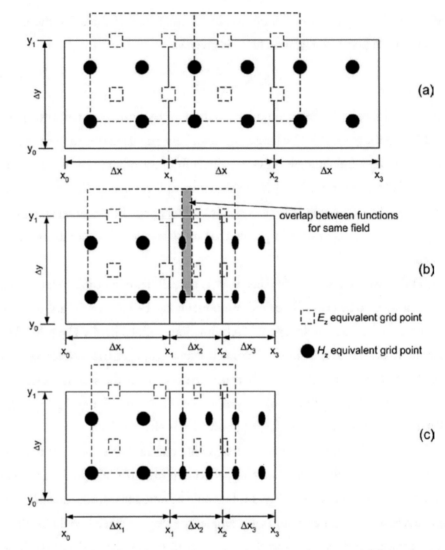

FIGURE 15: Offset between electric and magnetic fields in MRTD (a) fixed grid, (b) nonuniform grid (implemented incorrectly), and (c) nonuniform grid (implemented correctly).

points and the domain of the scaling functions are shown. The E_z field is offset in the z direction (normal to the page), while the H_z field is offset in the x and y directions.

Fig. 15(a) demonstrates a uniform MRTD grid. In this case, both scaling functions have dimensions $\Delta x \times \Delta y$. The scaling functions for each field start at the same point relative to the grid, and form a nonoverlapping basis covering the entire space. Fig. 15(b) demonstrates a nonuniform MRTD grid. In this case, the Δx_1 is twice Δx_2. The E_z field is not offset in the x direction, and thus for each cell $\Delta x = \Delta x_i$. The H_z field, however, is offset in the x direction. When $\Delta x = \Delta x_i$ is used for this field, the H_z field for the first cell continues to the center of the neighboring cell. The H_z field for the second cell, however, begins one quarter of the cell width from x_1. This causes the field coefficients that represent the H_z field to overlap. While it may be possible to use this field arrangement for MRTD, the overlap between neighboring field components would render the scheme implicit, and the wavelet basis nonorthogonal.

In order to keep the wavelet scheme nonoverlapping, and still allow for a variable grid, the width of the offset scaling functions must be set so that it begins s_d from the current grid point, and ends s from the next grid point (note that s_d is proportional to the cell spacing). This is demonstrated in Fig. 15(c). A scheme for determining the grid size for the different basis functions can then be quantified:

1. For nonoffset directions (directions normal to field component for electric fields, or direction of field component for magnetic fields),

$$\Delta n = \Delta n(l), \tag{180}$$

where l is an index denoting the position along the axis.

2. For offset directions (direction of field component for electric fields, or normal to field component for magnetic fields),

$$\Delta n = \Delta n(l) + \frac{\Delta n(l+1) - \Delta n(l)}{4}. \tag{181}$$

4.3   MRTD GRID EXCITATION

In Chapter 3, a method was presented that allows the addition of an impressed field to the MRTD field update equations. A modified $\mathbf{D_x}$ update was presented, (129), that allows a source function to be applied at specific equivalent grid points and then be transformed back into the wavelet domain. While this equation can be used to impress a field at any (or every) point in the MRTD grid, it is likely that only a relatively small number of grid points will be modified to excite the grid. In this case, the fields can be excited using

$$\mathbf{D}\mathbf{W}_{\text{dir1}} = \mathbf{D}\mathbf{W}_{\text{dir1}} + \mathbf{R}^{-1}\mathbf{J}, \tag{182}$$

after the fields are updated, if $\mathbf{J}$ is a vector of field values that is only nonzero at desired equivalent grid points. In this case, (129) is simply split into two steps.

One of the most common types of excitations for microwave circuit simulation is the microstrip line. One method of exciting a microstrip line is to impress a constant electric field in a plane normal to the microstrip metal, directly under the microstrip. For any wavelet resolution, it is necessary to use (182). This is because, when wavelets are applied, there is more than one equivalent grid point in each direction. This relatively simple excitation is demonstrated in Fig. 16.

In this example, $r_{\text{max}} = 0$ in all directions; therefore, there are two equivalent grid points per cell in each direction (eight total equivalent grid points). In this example, only the fields directly under the microstrip are excited. The field strength is represented by arrows, demonstrating that the impressed field at all points is identical in magnitude and direction. The arrows are located at the E_y equivalent grid points, and the cells are represented by the alternating shading. This is meant

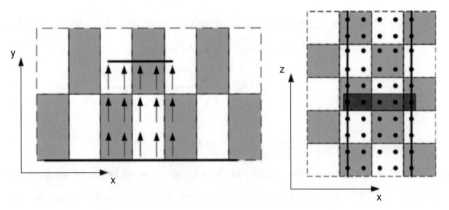

FIGURE 16: Excitation in MRTD, $r_{\text{max}} = 0$ cells demonstrated by alternate shading.

as simple approximation of the microstrip mode; as the field propagates through the grid it quickly matches the physical microstrip field pattern. In this example, five E_y field points fall under the microstrip line. As there are two of the field points in the y direction per cell, one cell is split in the x direction, where only half is excited. Similarly, all cells that contain the PEC are split in the y direction.

As in all other directions, there are two E_y field points in the z direction per cell. These field points are represented by the dots in Fig. 16, and the boundaries of the top metallization are the dark lines. For the excitation, however, only a single set of fields is excited. If both are excited, the fields resemble two identical, slightly offset pulses, needlessly complicating analysis. This is similar to subcell PEC analysis, where PECs can be simulated with an even number of cells in their tangential directions, but without subcell modeling, the PECs are several equivalent cells in thickness.

Similar limitations apply to coplanar waveguide (CPW) excitations; this case is presented in Fig. 17. For CPW excitations, the field is guided between the center

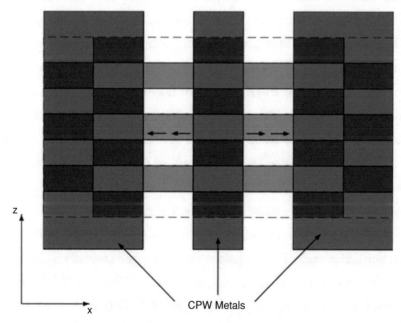

FIGURE 17: CPW excitation, demonstrating subcell excitation.

conductor and the surrounding ground planes. In this case, the field is excited as a voltage between the center conductor and the outer ground planes. In the direction of propagation (the z direction in the diagram) the field is only one cell thick. The impressed voltage along a line one equivalent cell in thickness can also used for wire antennas and probe feeds (in coaxial fed microstrip antennas or waveguides, for example).

4.4   TIME/SPACE ADAPTIVE GRIDDING

At the beginning of Chapter 3, the two main advantages of the MRTD method are presented. The first of these advantages is that many of the wavelet basis functions that can be used in MRTD allow the use of fewer basis functions per wavelength than FDTD, resulting in fewer coefficient updates per time step. However, it can be very difficult to apply PEC boundary conditions and the UPML boundary with these cases. In this book, the focus is placed on Haar wavelet basis functions, and while it is true that the Haar cells are larger than FDTD cells, the total number of scaling/wavelet coefficients needed per wavelength is the same as in the FDTD case. The advantage of the Haar scheme is the time-and-space adaptive grid.

The methods that have been presented allow for the modeling of general structures using Haar MRTD and are derived using an equivalence between FDTD and Haar MRTD. If the Haar MRTD method is applied, and the maximum wavelet resolution is used in all cells, the scheme is equivalent to FDTD. However, the chief advantage of the MRTD method is that the wavelet resolution can be varied from cell to cell. Most simulations are not constrained by the dispersion requirement, but instead by the structures that are simulated. For accurate simulation, several grid points must be used across each feature. Using Haar MRTD, the resolution can be locally increased to model complex structures, and reduced elsewhere for computational efficiency. This technique can be applied statically before the start of simulation, and automatically during simulation. When applied during simulation, this feature is usually referred to as the MRTD time-and-space adaptive grid [30].

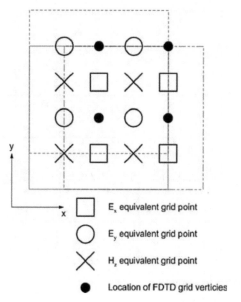

E_x equivalent grid point

E_y equivalent grid point

H_z equivalent grid point

Location of FDTD grid verticies

FIGURE 18: Haar MRTD cell (2D) showing FDTD grid points, $r_{max} = 0$.

The concept of the equivalent grid point was introduced in Chapter 2 to demonstrate the offset of the electric and magnetic fields in MRTD. In Haar MRTD the equivalent grid points are functionally identical to FDTD grid points (regions of constant field value). When wavelets are added and subtracted from a Haar representation, the effect is to add and remove equivalent grid points. This allows dense discretizations to be used in the area of complex structures, and coarse discretizations to be used elsewhere. Fig. 18 shows the arrangement of the fields (equivalent grid points) in a single $r_{max} = 0$ Haar MRTD cell, and also shows the positions of the FDTD grid points if the Yee-cell in Chapter 2 is used. These FDTD equivalent grid points can be used to demonstrate the FDTD equivalent grid for Haar MRTD adaptive resolution.

As an example of the capabilities of the Haar MRTD adaptive grid, consider the grid presented in Fig. 19. In this example, three wavelet resolutions are used from $r_{max} = -1$ (scaling function only) to $r_{max} = 1$. In the $r_{max} = -1$ case, there is one equivalent FDTD cell per MRTD cell; in the $r_{max} = 1$ case, there are eight equivalent FDTD cells per MRTD cell (for a 2-D example). Similar to Fig. 18, the

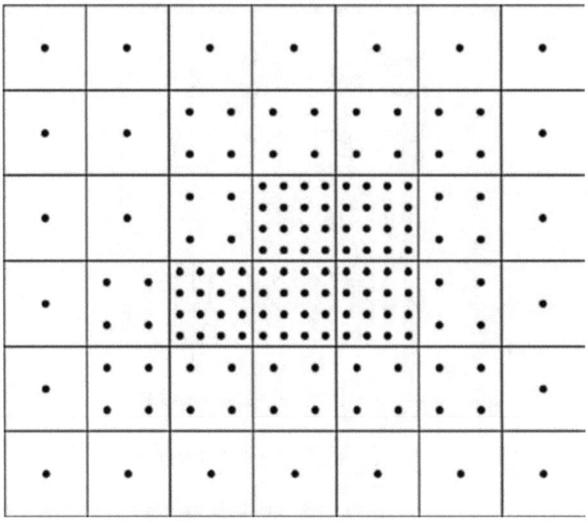

FIGURE 19: Adaptive grid example.

dots represent the equivalent FDTD cells, and the surrounding squares represent the MRTD grid boundaries.

In Fig. 19, the adaptive grid is used to place a dense discretization in the center, and a coarse discretization in the surrounding area. Using this grid, a highly detailed structure can be modeled in the center, while the surrounding, homogeneous area is simulated with low resolution. This technique is more powerful than the nonuniform mesh that is presented earlier in this chapter because the spacing of the equivalent grid points is a function of all three grid dimensions, and thus can be increased locally to represent complex features. One important note is that, while this example demonstrates the grid when the resolution is the same in all dimensions, this is not a requirement.

While the grid that is presented in Fig. 19 can be fixed for the entire simulation, the MRTD method also provides the ability to change the resolution during simulation. In Chapter 3, it is demonstrated that at any point in the grid, the highest resolution wavelet represents the deviation from the average. If the field is not changing rapidly, the high resolution wavelets can be neglected during the simulation. In [30], it is suggested that an absolute and relative threshold be used to determine whether a wavelet is needed. This concept forms the basis for the adaptive grid.

For the examples presented in the next chapter, the thresholds were chosen as in [30]. The relative threshold is a fraction of the value of the scaling function. If the value of the wavelet is less than this fraction of the scaling function, it can be neglected. The absolute threshold is required for low-field values, and does not change during simulation. If the fields are near zero, the difference between the scaling and wavelet coefficients will not be large. However, the wavelets are not required because the field does not vary rapidly. In this case, the wavelets will be lower than the absolute threshold, and can be neglected.

This scheme is not effective if it is applied every time step. For the test to be applied, the wavelet value must be calculated, which defeats the purpose of the adaptive grid. In this work a scheme was used where the wavelets were tested at a user defined period. Wavelets that are below either threshold are neglected until the next testing period. Using this scheme, only the highest level wavelets are tested. If the wavelets are removed, at the next testing period both the currently highest level wavelets are tested (one lower than the previous level) and the next highest level is tested. In this manner, it can be determined if wavelets need to be reintroduced into the grid.

In the code that was used in this simulation, both time-adaptive grids and static (but variable in space) grids are used. Using this scheme, high-resolution grids can be used to specify complex structures, while the resolution around the structure can be predefined to a low value. In this manner, the adaptive grid can be applied to vary the resolution in response to field propagation, while the resolution is not higher than necessary at any point in the structure.

CHAPTER 5

MRTD Simulation Examples

The techniques that are presented in this book can be used to simulate a wide variety of structures in MRTD. Using the subcell modeling of arbitrary structures, any structure that can be simulated in FDTD can be simulated in MRTD. If a structure can be represented within a cell, then the cells used in an MRTD simulation can be significantly larger. Using the method presented here, any structure that can be simulated in FDTD can be simulated with MRTD. In addition, the MRTD time-and-space adaptive grid can be applied to allow fewer grid points in the MRTD case.

In order to verify the method, several simulations were run using the MRTD code that was developed. The code utilizes all of the features that have been discussed, including the UPML, nonuniform gridding, and lumped element modeling. For the practical cases, the MRTD results were verified by an FDTD code.

To measure the effectiveness of the MRTD code, both simulation time and number of equivalent grid points required to simulate the structure were recorded. The number of equivalent grid points, which is equal to the total number of grid points in the FDTD case, is useful because it directly relates to the number of calculations required. This is the best measure of how one MRTD grid compares to another. In the cases where the time-adaptive grid was used, the number of equivalent grid points provided a method to measure the effectiveness of the adaptive scheme.

5.1   ANALYZING MRTD OUTPUT

The microstrip structures that were evaluated in the following examples were measured by computing the voltage on the microstrip lines. The voltage is calculated using

$$V_{ab} = -\int_{ab} \mathbf{E}(\mathbf{r}) \cdot \partial \mathbf{l}, \tag{183}$$

where a is the ground plane and b is the microstrip conductor. This is the same method that is used for FDTD simulations [14]. Similarly, current is calculated using

$$I = \oint_{C} \mathbf{H}(\mathbf{r}) \cdot \partial \mathbf{l}, \tag{184}$$

where C is the contour that the current passes through. These are relatively simple calculations, and are solved in the discrete case by summing the electric or magnetic field values along the path and multiplying by the space step. Of course, in MRTD, the field must be reconstructed to perform the sum, and the space step used must be the equivalent cell size. In this case it is useful to note that while the reconstruction is a matrix operation, the field at any one point in the cell can be reconstructed by multiplying the coefficient vector by the row of the reconstruction matrix that represents the desired point.

Most structures are not characterized in the frequency domain; therefore, the time domain voltages must be transformed into the frequency domain using a discrete Fourier transform (any DFT method will suffice). Once the frequency domain voltages are determined, they can be used to calculate device parameters such as S parameters and characteristic impedance. S parameters are calculated using [14]

$$S_{mn} = \frac{V_m(\omega)}{V_n(\omega)} \sqrt{\frac{Z_{0,n}(\omega)}{Z_{0,m}(\omega)}}, \tag{185}$$

where the V_m is the voltage out of port m and V_n is the voltage into port n. In most simulations, one port is excited and the output is measured at the other ports. In this case the total voltage measured at the output ports can be used directly in (185). The voltage recorded at the input, however, is the total of the input waveform and reflections from the structure being characterized. In this case, a separate simulation of the input line alone can be used to determine the input. This "thru" case can be used as V_n in (185). For the special case of S_{nn}, the following can be used:

$$S_{nn} = \frac{V_{n,total}(\omega) - V_{n,thru}(\omega)}{V_{n,thru}(\omega)}. \tag{186}$$

For the microstrip lines used here, the characteristic impedance can be calculated simply by dividing the voltage between the ground and the conductor at one point with the current through the conductor at the same point. However, the electric and magnetic fields are offset. Both the half-time step and half equivalent grid point offset must be accounted for. Using [31]

$$Z_o(\omega) = \frac{V_z(\omega) e^{-j\omega\Delta t/2}}{\sqrt{I_{z-1/2}(\omega) I_{z+1/2}(\omega)}}, \tag{187}$$

the characteristic impedance can be determined. In this case, the line runs in the z direction and the $z - 1/2$ and $z + 1/2$ subscripts indicate that the current values are offset from the voltage values by one half of a space step.

5.2   PML ABSORPTION USING MICROSTRIP LINE

The first test that was run of the MRTD code was that of a simple microstrip line. Beyond a simple check to see if the code is working, this test allows the performance of the PML to be evaluated. As has already been presented, the PML is a numerical representation of a nonphysical (at least at the time of this printing) high-loss material that is perfectly matched for incoming waves at any angle. When implemented numerically, however, the material exhibits small reflections. The reflections are generated in two ways. First, the medium is highly lossy, but not infinitely lossy; some of the incoming wave will propagate through the material, reflect off of the outer PEC boundary, and reenter the simulation space. In addition, when implemented in a discrete space, the PML is not perfectly matched and small reflections occur. The parameters that can be used to finely tune the PML are σ, k, and the degree of the polynomial that is used to grade them. Using these parameters, it is possible to characterize the PML for a number of applications and values of the parameters [14].

The purpose of the test used here was not to completely characterize the PML, but rather to verify that the PML was functioning as expected. The test structure is a microstrip line terminated at one end with PML. The microstrip line is excited using the method presented in the previous chapter. The time shape of the pulse is Gaussian, with time duration chosen such that the 3 dB point is at 30 GHz. Two structures are simulated. In the first case, the line is chosen to be sufficiently long so that reflections from the far wall do not return until the initial pulse (and its reflection from the near wall) has passed. In the second case, the line is terminated with PML very close to the excitation.

The simulation was performed using $r_{max} = 1$ in all directions (64 equivalent grid points per cell). The microstrip PEC is 452-μm wide and the substrate is 200-μm thick with $\varepsilon_r = 2.2$. The predicted Z_0 is 50 Ω. The grid size is 452 μm across the microstrip line, so that four equivalent grid points are used to model the line. The cell height is 200 μm; the entire height of the substrate is modeled with one cell (four equivalent grid points). Using this grid, subcell PEC modeling is necessary to restrict the microstrip conductor to only one cell in height.

For this case, S_{11} was calculated to quantify the reflection from the PML. The reflection, as seen in Fig. 20, is lower that −60 dB at 10 GHz, and slowly increases with frequency. This is expected behavior; the reflection from the PML is caused primarily by the effect of the discrete grid. As the frequency increases, the relative size of the cells also increases. The PML could most likely be tuned by manipulating σ, k, their variation across the PML, and the grid size, but it is not required for the structures that are characterized in the further examples. The response of the featured antenna is significantly higher than this noise floor, making the reflection irrelevant.

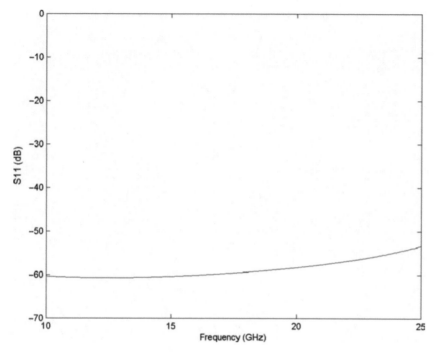

FIGURE 20: Reflection from PML.

5.3   LUMPED ELEMENT VERIFICATION: RESISTOR-TERMINATED MICROSTRIP LINE

In order to test the lumped element algorithm that is presented in Chapter 3, a very simple case was simulated. A microstrip line, the same as in the antenna example, was terminated with resistors. This technique is commonly used to simulate measurement environments. As was shown in the previous example, the PML has low reflections across a wide band. In measurement, the ports are often terminated with a load of fixed impedance. If the input line is not matched to the load, or the characteristic impedance of the line changes with frequency, then reflections will occur from the termination. By terminating the simulation with resistors, the measurement environment is simulated.

In this example, the microstrip line of the previous example is terminated with a PML at one end and five parallel resistors at the other. The PEC that represents the microstrip conductor is four equivalent grid points wide (5 E_z field locations).

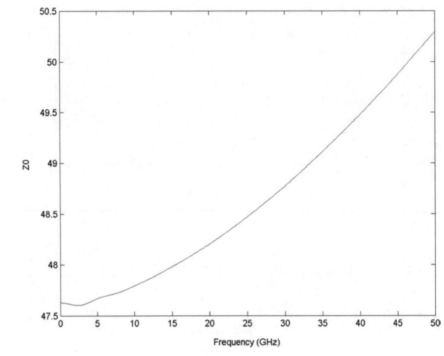

FIGURE 21: Characteristic impedance of microstrip line.

The impedance of the line was first calculated using (188), and is presented in Fig. 21. The impedance can be seen to vary slightly over the 50 GHz band. The resistors were chosen to represent a 49 Ω load (each resistor is 245 Ω), the impedance at the center of the band.

The time domain response is given in Fig. 22. The time domain is presented instead of the S parameters because, in this case, the reflection from the resistor can be seen. The initial pulse is a Gaussian derivative, with the 3 dB frequency at 50 GHz. The reflection occurs at approximately 175 ps, and is the small ripple in the response. In the case where both ends of the line are terminated with PML, the ripple does not appear. This response demonstrates that the resistor is matched for part of the band, but not all. This is verified by S_{11}, which is presented in Fig. 23, demonstrating that the best match occurs at approximately 37 GHz, the same frequency that $Z_0 = 49\Omega$.

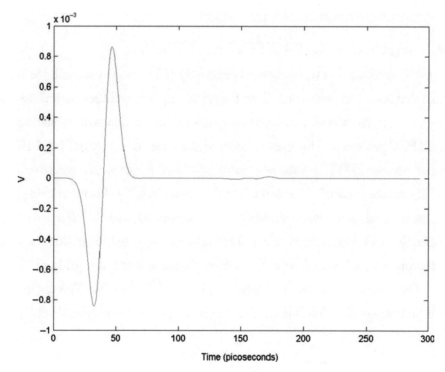

FIGURE 22: Time domain reflection from resistor.

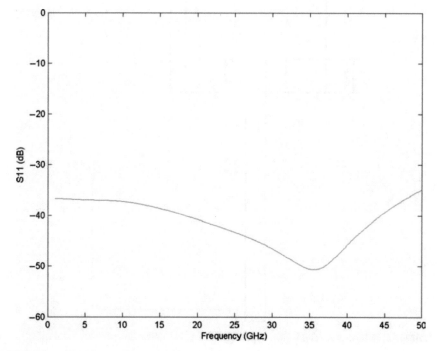

FIGURE 23: Frequency domain reflection from resistor.

5.4   MICROSTRIP PATCH ANTENNA

The PML example that was presented effectively validates the code and the techniques that have been presented. Both the microstrip line and the resistors used to terminate it in the previous example are smaller than a cell in size, validating the subcell PEC procedure. However, in order to show the advantage of Haar MRTD simulations over FDTD simulations (and verify that the code gives results similar to FDTD), another example is presented. In this example, a 30 GHz microstrip patch antenna is simulated. Several simulations are performed, and the effectiveness of the adaptive grid over a variety of threshold values is evaluated. In addition, a static multiresolution grid is used, as well as a mix of static and adaptive grids.

The antenna that was simulated is presented in Fig. 24. The antenna is based on a design for a 30GHz patch antenna built on a thin polyimide film [32].

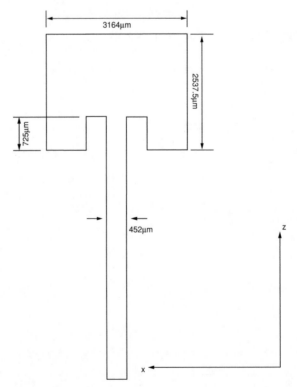

FIGURE 24: Microstrip fed patch antenna.

The microstrip line is identical to the line used in the previous two examples. In the simulation, $r_{max} = 1$ is used in all directions. A uniform grid is used with $\Delta x = 452\,\mu m$, $\Delta y = 200\,\mu m$, and $\Delta z = 725\,\mu m$. The uniform grid is used to ensure that the FDTD grid and the MRTD grid are identical. When a nonuniform grid is used in FDTD, a change in cell size affects only adjacent cells. However, in the MRTD case, the change in cell size affects every equivalent cell within the MRTD cell. Using the offset method presented here, an FDTD grid cannot be generated from the MRTD grid by simply subdividing the grid size. While it is possible to match an MRTD and FDTD grid (by applying a variable grid to every cell in the FDTD grid), the purpose of this example is to measure the effect of the MRTD adaptive grid.

To avoid interaction of the antenna with the boundary, a space of three times the antenna width is placed between the PML and the antenna on all sides (three times the substrate height on the top). Including the antenna, the surrounding space, and the absorber, the simulation space is 40 cells × 7 cells × 23 cells. In equivalent grid points (and FDTD cells) the grid is 160 × 28 × 92. To model the antenna, a 50 GHz Gaussian derivative pulse is used to excite the microstrip line. As a reference to the adaptive grid case, an FDTD simulation and MRTD simulation with constant grid were run. The time domain voltage on the feed line is presented in Fig. 25.

In this figure, excellent agreement is observed between the MRTD and FDTD simulations. However, there are small differences between these time domain plots and these are revealed in the plot of S_{11}, Fig. 26. It can be seen from the S_{11} plot that there is a small difference between the simulations (approx 1.5%) in the resonance frequency of the antenna. When the time response is examined more closely, Fig. 27, the differences between the FDTD and MRTD simulations are easier to observe. The primary reason for this difference is the modeling of the varying dielectric in the different techniques. In the FDTD simulation, the ε_r value used in the update equation is applied locally. For field values on the air/substrate interface, the average of the two values is used. In the MRTD case, the discretization

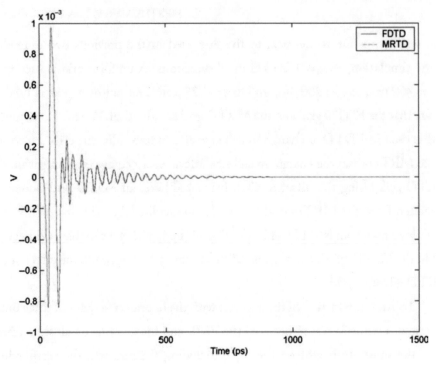

FIGURE 25: Time domain response of antenna, FDTD and MRTD, $r_{max} = 1$.

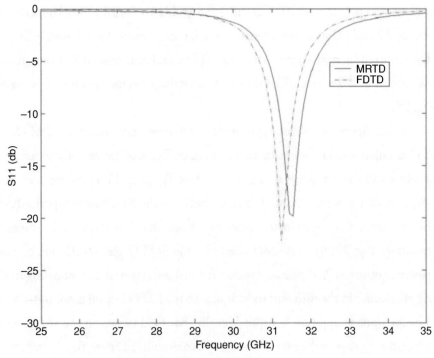

FIGURE 26: S_{11} of patch antenna, MRTD/FDTD comparison.

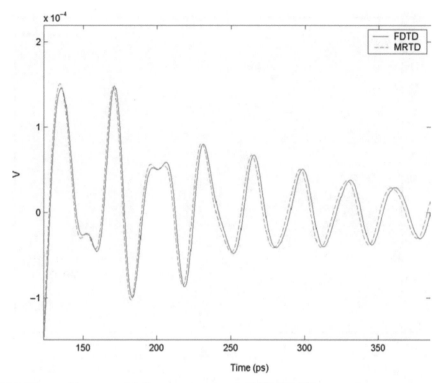

FIGURE 27: Close-up of differences between MRTD/FDTD, time domain.

of the constitutive relationship involves field values across the entire cell, leading to the small difference between the two techniques.

To verify that the differences in the methods that are used to model material interfaces is responsible for the frequency shift in Fig. 26, another experiment was run where ε_r was kept constant across the entire space. For this experiment, the ε_{eff} (2.47) of the microstrip line was chosen. For this case, S_{11} is presented in Fig. 28. In this example, the MRTD and FDTD results overlap; there is no frequency shift.

The remainder of the cases that were simulated using this technique are experiments with the grid resolution. The first tests that were run used a static grid, but lowered the resolution in the area surrounding the antenna. This is a very important test because it demonstrates that the technique can be used to locally increase the grid resolution to simulate a complex structure (the antenna), while the

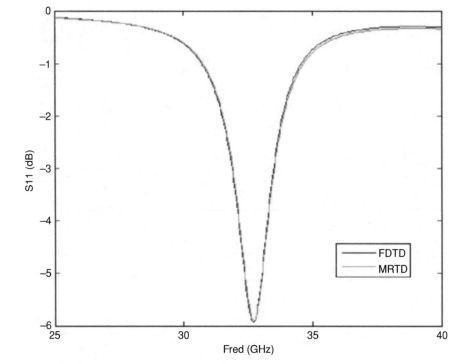

FIGURE 28: FDTD/MRTD comparison, constant ε_r.

surrounding area can use a coarser grid. In this case, the smallest grid dimension used in the x direction (tangential to the antenna surface) is $\lambda/53.1$. The grid is sampled significantly higher than required for accuracy in free space; however, this grid is necessary to accurately model the feed. When the resolution is reduced in the area surrounding the antenna, the spacing is $\lambda/26.5$ for $r = 0$ and $\lambda/13.3$ for $r = -1$ (scaling function only).

The results for the $r = -1$ case are presented in Fig. 29. The grid in this case has 12.4% fewer grid points than in the grid that uses the same resolution everywhere. If the PML cells are neglected for this calculation, the number of cells is reduced by 34%. This modest improvement is due to the relatively small structure that is simulated. In this structure, the only area that can be reduced in resolution is the area surrounding the antenna, and it uses relatively small number of cells. In this case, the grid has 6440 cells (with 64 equivalent grid points per cell for a total

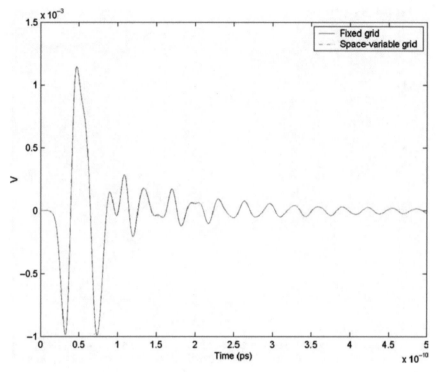

FIGURE 29: Comparison of $r_{\max} = 1$ MRTD with surround grid fixed at $r = -1$.

of 412,160 equivalent grid points), and 1077 of these cells surround the antenna and are reduced in resolution (cells in the PML are not reduced). In addition, the cells on the bottom of the grid intersect the ground plane, and the resolution is not lowered in the y direction. The results for the $r = 0$ case are presented in Fig. 30, the number of non-PML grid points is reduced by 28.1%.

The remaining cases that are presented demonstrate the adaptive grid. In these cases the resolution was set to one at the beginning of the simulation, and relative and adaptive thresholds, t_r and t_a respectively, were used to in the manner indicated in the previous chapter to automatically determine if wavelets were needed. The t_a values that are presented are relative to the maximum value of the pulse for the entire simulation (this can be determined *a-priori* from the excitation). A number of cases were run to test the technique; representative cases are presented

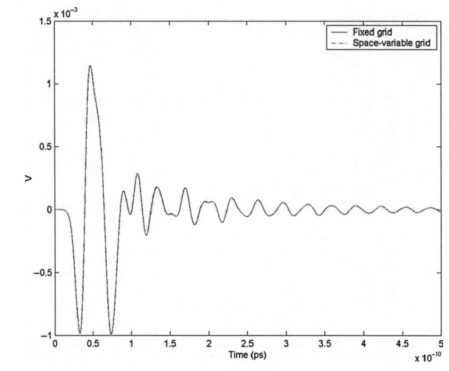

FIGURE 30: Comparison of $r_{max} = 1$ MRTD with surround grid fixed at $r = 0$.

here. In all of these cases, the thresholding is performed every 110 time steps. This value is chosen because it represents the Nyquist limit for the resonant frequency of the antenna. When the thresholding is performed significantly more often than this rate, the effectiveness is largely unchanged, and for values higher than this limit, the scheme is slightly less efficient.

In the first two cases that are presented, the threshold values are too high. In Fig. 31, the simulation uses $t_r = 0.1$ $t_a = 0.01$. In the time domain output, it can be seen that the oscillations of the wave are quickly damped. This is because the adaptive grid algorithm removes wavelets that have significant values. A similar result is presented in Fig. 32. In this case, both thresholds are reduced by an order of magnitude from the previous case. It can be seen that the damping takes a longer time to occur. It is interesting to note that in both of these cases the results at the time before the results are abruptly damped closely match the fixed grid case.

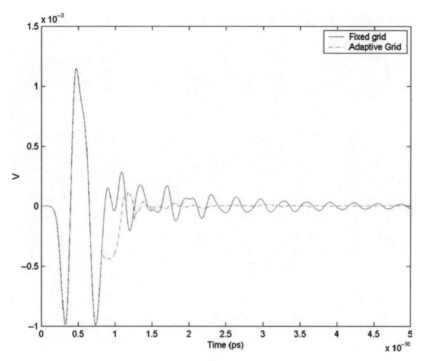

FIGURE 31: Comparison of fixed grid MRTD with adaptive grid $t_r = 0.1$ and $t_a = 0.01$.

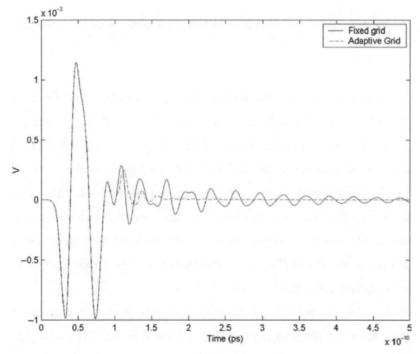

FIGURE 32: Comparison of fixed grid MRTD with adaptive grid $t_r = 0.01$ and $t_a = 0.001$.

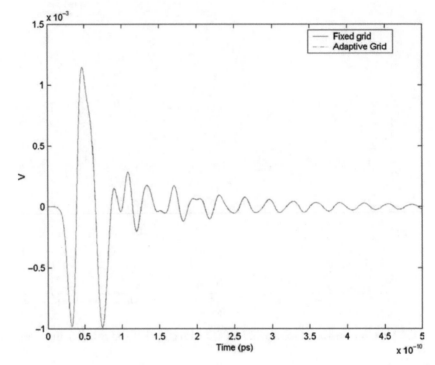

FIGURE 33: Comparison of fixed grid MRTD with adaptive grid $t_r = 1 \times 10^{-4}$ and $t_a = 1 \times 10^{-5}$.

In the final case that is presented the results are very close to the fixed grid case (0.73% error). The error in this case was calculated by integrating the magnitude of the difference between the uniform and variable grid cases and dividing by the integral of the magnitude of the uniform case. In Fig. 33, the simulation uses $t_r = 1 \times 10^{-4}$ and $t_a = 1 \times 10^{-5}$. In this case the adaptive algorithm does not remove wavelets that are needed. However, results exactly the same as the fixed grid case could easily be achieved by setting the thresholds to an extremely low value. For the method to be effective, it must also reduce the number of calculations required to update the fields.

In Fig. 34, the number of the equivalent grid points, normalized to the maximum, at each time step is presented for the $t_r = 1 \times 10^{-4}$, $t_a = 1 \times 10^{-5}$ case. This plot can be used to calculate the total number of grid points required

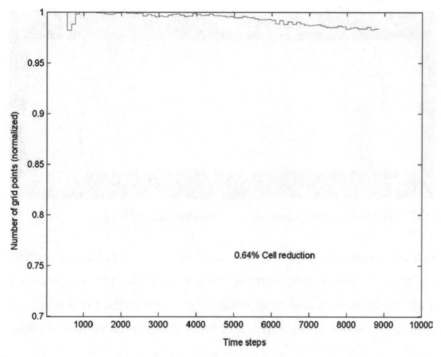

FIGURE 34: Number of gridpoints used in simulation vs. time, $t_r = 1 \times 10^{-4}$ and $t_a = 1 \times 10^{-5}$.

for the simulation. For the case that is presented, the reduction in grid points is only 0.6%. This is extremely low, and is most likely because the structure being simulated occupies most of the grid. The adaptive algorithm will most likely yield higher efficiency in grids with finely detailed structures that are dispersed in a larger grid. In the next section, one of these structures is explored.

5.5   DUAL MICROSTRIP PATCH ANTENNAS

The previous example served to demonstrate the subcell technique as applied to a relatively simple structure and showed the benefit of both fixed subcell grids and adaptive grids. The adaptive gridding method that has been presented allows the grid to be varied as a function of position and provides modest gains for many structures. This technique is even more effective for highly detailed structures with

FIGURE 35: Dual microstrip patch antenna (shaded area is PML).

large homogeneous areas. For this example, the antenna from the previous section is employed. Another similar antenna, this one tuned for 43 GHz, is placed on the same substrate, separated by approximately three wavelengths. An example of a situation where this configuration may be used is in a device that has separate send and receive bands. In this case, it is useful to know the isolation between the antennas and the change in the radiation due to the proximity of the antennas. This structure is presented in Fig. 35.

For this example, the feed for the 31GHz antenna is labeled port one, and the feed for the 43GHz antenna is labeled port two. For this structure, S_{11} and S_{22} are presented in Figs. 36 and 37, respectively. The return loss for the 31 GHz antenna is almost exactly the same as in the case where the antenna was evaluated alone. This can be further explained by examining S_{21}, the coupling between the antennas. In this case, the maximum coupling occurs, as expected, at 43 GHz (the resonant frequency of the second antenna), but it is very low; the isolation between the antennas is 48 dB.

This simulation was evaluated with the same metrics as the previous example. A fixed subcell grid was employed, and the difference between the two simulations was examined. For the fixed grid, high resolution was used to model the antenna, while low resolution was used around the antenna. The antenna was simulated with $r = 1$, and the surrounding area used $r = -1$ (scaling function only). It is important

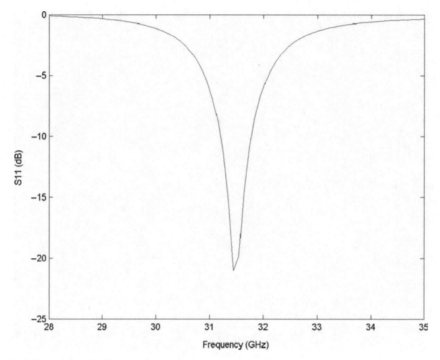

FIGURE 36: S_{11} for dual antennas.

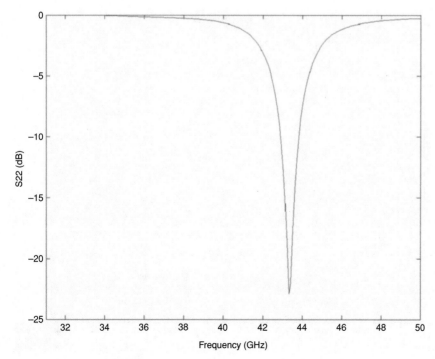

FIGURE 37: S_{22} for dual antennas.

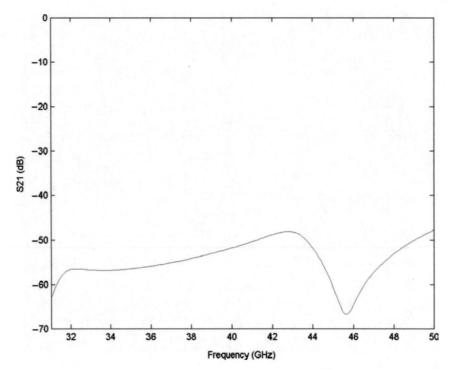

FIGURE 38: S_{21} for dual antennas.

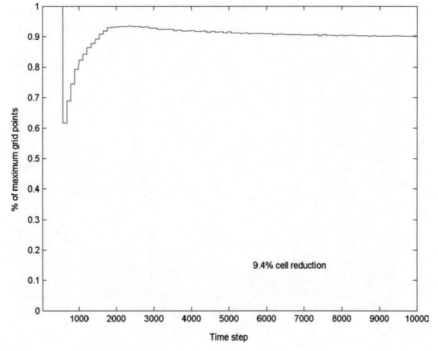

FIGURE 39: Adaptive grid for initial $r = 1$ everywhere.

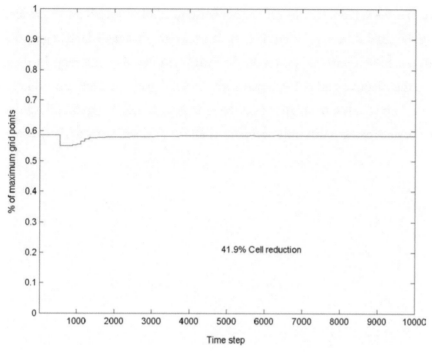

FIGURE 40: Adaptive grid with resolution surrounding antenna fixed at $r = -1$.

to note that even in the $r = -1$ case, there are still more than 12 equivalent grid points per wavelength. In the previous example, it was noted that the fixed subcell grid used less overall grid points than the adaptive grid case (where the resolution at the beginning was $r = 1$ everywhere). In this example, a simulation was run where the adaptive grid was used in the area of the antenna, while the surrounding resolution was still fixed at $r = -1$. The goal of this simulation was to determine what improvement, if any, could be obtained through the application of both the adaptive and fixed subcell grids.

For the fixed subcell grid case, where the resolution between and surrounding the antennas was reduced to -1, the efficiency when compared to the uniform grid case is 41.4%. The two other cases of interest are the adaptive grid for the entire grid and the adaptive grid for the antenna area only (with the surrounding grid at -1 as in the fixed subcell case). For the case where only adaptive gridding is used, there is a 9.4% reduction in cells. This is significantly lower than the reduction using the fixed subcell grid, but it does demonstrate that the adaptive grid can be used to achieve nontrivial grid reduction while maintaining accuracy. In the case where the adaptive grid is combined with the fixed grid, the total cell reduction is 41.9%. This is only 0.6% higher than the fixed grid case, showing that the adaptive grid does not cause much improvement when the grid is already fairly sparse.

References

[1] K. S. Yee, "Numerical solution of initial boundary value problems involving Maxwell's equations in isotropic media," *IEEE Trans. Antennas Propagat.*, vol. AP-14, pp. 302–307, Mar. 1966. doi:10.1109/TAP.1966.1138693

[2] M. Krumpholz and L. P. B. Katehi, "New time domain schemes based on multiresolution analysis," *IEEE Trans. Microwave Theory Tech.*, vol. 44, pp. 555–561, April 1996. doi:10.1109/22.491023

[3] M. Fujii and W. J. R. Hoefer, "A three-dimensional Haar-wavelet-based multiresolution analysis similar to the FDTD method—derivation and application," *IEEE Trans. Microwave Theory Tech.*, vol. 46, pp. 2463–2475, Dec. 1998. doi:10.1109/22.739236

[4] Y. Tretiakov and G. W. Pan, "On Daubechies wavelet based time domain scheme," *Proc. Antennas Prop. Sym.*, vol. 4, pp. 810–813, July 2001.

[5] T. Dogaru and L. Carin, "Multiresolution time-domain using CDF biorthogonal wavelets," *IEEE Trans. Microwave Theory Tech.*, vol. 49, pp. 902–912, May 2001. doi:10.1109/22.920147

[6] I. Daubechies, *10 Lectures on Wavelets*, Society for Industrial and Applied Mathematics, Philadelphia, 1992.

[7] G. Carat, R. Gillard, J. Citerne, and J. Wiart, "An efficient analysis of planar microwave circuits using a DWT-based Haar MRTD scheme," *IEEE Trans. Microwave Theory Tech.*, vol. 48, pp. 2261–2270, Dec. 2000. doi:10.1109/22.898973

[8] T. Dogaru and L. Carin, "Application of Haar-wavelet-based multiresolution time-domain schemes to electromagnetic scattering problems," *IEEE Trans. Antennas Propagat.*, vol. 50, no. 6, pp. 774–784, June 2002. doi:10.1109/TAP.2002.1017657

[9]   C. Sarris and L. P. B. Katehi, "Fundamental gridding-related dispersion effects in multiresolution time-domain schemes," *IEEE Trans. Microwave Theory Tech.*, vol. 49, no. 12, pp. 2248–2257, Dec. 2001. doi:10.1109/22.971607

[10] Z. Chen and J. Zhang, "An unconditionally stable 3-D ADI-MRTD method free of the CFL stability condition," *IEEE Microwave Wireless Comp. Lett.*, vol. 11, pp. 349–351, Aug. 2001. doi:10.1109/7260.941786

[11] F. Zheng, Z. Chen and J. Zhang, "Toward the development of a three dimensional unconditionally stable finite-difference time-domain method," *IEEE Trans. Microwave Theory Tech.*, vol. 48, pp. 1550–1558, Sept. 2000. doi:10.1109/22.869007

[12] E. Tentzeris, R. L. Robertson, J. F. Harvey, and L. P. B. Katehi, "Stability and dispersion analysis of Battle-Lemarie-Based MRTD schemes," *IEEE Trans. Microwave Theory Tech.*, vol. 47, pp. 1004–1013, July 1999. doi:10.1109/22.775432

[13] A. Taflove and M. E. Brodwin, "Numerical solution for steady-state electromagnetic scattering problems using the time-dependent Maxwell's equations," *IEEE Trans. Microwave Theory Tech.*, vol. 23, pp. 623–630, Aug. 1975. doi:10.1109/TMTT.1975.1128640

[14] A. Taflove and S. Hagness, *Computational Electromagnetics, the Finite-Difference Time-Domain Technique*, 2nd ed. Artech House, Boston, 2003.

[15] R. Robertson, E. Tentzeris, M. Krumpholz, and L. P. B. Katehi, "MRTD analysis of dielectric cavity structures," *Proc. IEEE MTT-S*, vol. 3, pp. 1861–1864, June 1996.

[16] E. Tentzeris, M. Krumpholz, and L. P. B. Katehi, "Application of MRTD to printed transmission lines," *Proc. IEEE MTT-S*, vol. 2, pp. 573–576, June 1996.

[17] Q. Cao, Y. Chen, and R. Mittra, "Multiple image technique (MIT) and anistropic perfectly matched layer (APML) in implementation of MRTD scheme for boundary truncations of microwave structures,"

IEEE Trans. Microwave Theory Tech., vol. 50, pp. 1578–1589, 2002. doi:10.1109/TMTT.2002.1006420

[18] N. Bushyager, J. Papapolymeroy, and M. M. Tentzeris, "A composite-cell multiresolution time-domain technique for the design of antenna systems including electromagnetic band gap and via-array structures," *IEEE Trans. Antennas Prop.*, vol. 53, pp. 2700–2710, Aug. 2005. doi:10.1109/TAP.2005.851832

[19] Y. W. Cheong et al., "Wavelet-Galerkin scheme of time-dependent inhomogeneous electromagnetic problems," *IEEE Microwave Guided Wave Lett.*, vol. 9, pp. 297–299, Aug. 1999. doi:10.1109/75.779907

[20] M. Fujii and W. J. R. Hoefer, "Dispersion of time domain wavelet galerkin method based on daubechies' compactly supported scaling functions with three and four vanishing moments," *IEEE Microwave Guided Wave Lett.*, vol. 10, pp. 125–127, April 2000. doi:10.1109/75.846920

[21] K. R. Umashankar, A. Taflove, and B. Becker, "Calculation and experimental validation of induced currents on coupled wires in an arbitrary shaped cavity," *IEEE Trans. Antennas Propagat.*, vol. 35, pp. 1248–1257, Nov. 1987. doi:10.1109/TAP.1987.1144000

[22] A. Taflove et al., "Detailed FDTD analysis of electromagnetic fields penetrating narrow slots and lapped joints in thick conducting screens," *IEEE Trans. Antennas Propagat.*, vol. 36, pp. 247–257, Feb. 1988. doi:10.1109/8.1102

[23] S. Dey and R. Mittra, "A modified locally-conformal finite-difference time-domain algorithm for modeling three-dimensional perfectly conducting objects," *Microwave Optical Tech. Lett.*, vol. 17, pp. 349–352, April 1998. doi:10.1002/(SICI)1098-2760(19980420)17:6<349::AID-MOP4>3.0.CO;2-H

[24] J. G. Maloney and G. S. Smith, "The efficient modeling of thein material sheets in the FDTD method," *IEEE Trans. Antennas Propagat.*, vol. 40, pp. 323–330, March 1992. doi:10.1109/8.135475

[25] J. G. Maloney, G. S. Smith, and W. R. Scott, Jr. "Accurate computation of the radiation from simple antennas using the finite-difference time-domain method," *IEEE Trans. Antennas Propagat.*, vol. 38, pp. 1059–1068, July 1990. doi:10.1109/8.55618

[26] J. G. Maloney and G. S. Smith, "The use of surface impedance concepts in the finite-difference time-domain method," *IEEE Trans. Antennas Propagat.*, vol. 40, pp. 38–48, Jan. 1992. doi:10.1109/8.123351

[27] N. Bushyager, E. Dalton, and E. M. Tentzeris, "Modeling of complex RF/wireless structures using computationally optimized time-domain techniques," *Int. J. Num. Modeling*, vol. 17, no. 3, pp. 223–236, April 2004. doi:10.1002/jnm.542

[28] M. J. Piket-May, A. Taflove, and J. Baron, "FDTD modeling of digital signal propagation in 3-D circuits with passive and active loads," *IEEE Trans. Microwave Theory Tech.*, vol. 42, pp. 1514–1523, Aug. 1994. doi:10.1109/22.297814

[29] P. Monk and E. Suli, "Error estimates for Yee's method on non-uniform grids," *IEEE Trans. Magnetics*, vol. 30, pp. 3200–3203, Sept. 1994. doi:10.1109/20.312618

[30] E. M. Tentzeris, A. Cangellaris, L. P. B. Katehi, and J. Harvey, "Multiresolution time-domain (MRTD) adaptive schemes using arbitrary resolutions of wavelets," *IEEE Trans. Microwave Theory Tech.*, vol. 50, pp. 501–516, Feb. 2002. doi:10.1109/22.982230

[31] J. Fang and D. Xue, "Numerical errors in the computation of impedances by FDTD method and ways to eliminate them," *IEEE Microwae Guided Wave Lett.*, vol. 5, pp. 6–8, Jan. 1995. doi:10.1109/75.382377

[32] E. Y. Tsai, A. M. Bacon, M. Tentzeris, and J. Papapolymerou, "Design and development of novel micromachined patch antennas for wireless applications" in *Proc. Asian-Pacific Microwave Symposium*, Nov. 2002, pp. 821–824.

Biographies

Dr. Nathan A. Bushyager received a B.S. in Engineering Science from The Pennsylvania State University in 1999 and an M.S. and Ph.D. from The Georgia Institute of Technology in 2003 and 2004, respectively. In addition to this book he has authored a book chapter, six journal papers, and has presented more than 30 conference papers. In 2001, he received the Best Student Paper award at the Applied Computational Electromagnetics Society symposium. Dr. Bushyager is currently a Senior RF Engineer at Northrop Grumman in Linthicum Maryland.

Professor Manos M. Tentzeris received the Diploma Degree in Electrical and Computer Engineering from the National Technical University of Athens ("Magna Cum Laude") in Greece and the M.S. and Ph.D. degrees in Electrical Engineering and Computer Science from the University of Michigan, Ann Arbor, MI and he is currently an Associate Professor with School of ECE, Georgia Tech, Atlanta, GA. He has published more than 200 papers in refereed Journals and Conference Proceedings and 8 book chapters and he is in the process of writing 3 books. Dr. Tentzeris has helped develop academic programs in Highly Integrated/Multilayer Packaging for RF and Wireless Applications, Microwave MEM's, SOP-integrated antennas and Adaptive Numerical Electromagnetics (FDTD, MultiResolution Algorithms) and heads the ATHENA research group (15 researchers). He is the Georgia Tech NSF-Packaging Research Center Associate Director for RF Research and the RF Alliance Leader. He is also the leader of the RFID Research Group of the Georgia Electronic Design Center (GEDC) of the State of Georgia. He was the recipient of the 2006 IEEE MTT Outstanding Young Engineer Award, the 2004 IEEE Transactions on Advanced Packaging Commendable Paper Award, the 2003 NASA Godfrey "Art" Anzic Collaborative Distinguished Publication Award for his activities in the area of finite-ground low-loss low-crosstalk coplanar waveguides, the 2003 IBC International Educator of the Year Award, the 2003 IEEE CPMT Outstanding Young Engineer Award for his work on 3D multilayer integrated RF modules, the 2002 International Conference on Microwave and Millimeter-Wave Technology Best Paper Award (Beijing, CHINA) for his work on Compact/SOP-integrated RF components for low-cost high-performance wireless front-ends, the 2002 Georgia Tech-ECE Outstanding Junior Faculty Award, the 2001 ACES Conference Best Paper Award and the 2000 NSF CAREER Award for his work on the development of MRTD technique that allows for the system-level simulation of RF integrated modules and the 1997 Best Paper Award of the International Hybrid Microelectronics and Packaging Society for the development of design rules for low-crosstalk finite-ground embedded transmission lines. He was also the 1999 Technical Program Co-Chair of the 54th ARFTG Conference, Atlanta, GA and the Chair of the 2005 IEEE CEM-TD Workshop and he is the Vice-Chair of the RF Technical Committee (TC16) of the IEEE CPMT Society.

He has organized various sessions and workshops on RF/Wireless Packaging and Integration in IEEE ECTC, IMS and APS Symposia in all of which he is a member of the Technical Program Committee in the area of "Components and RF". He is the Associate Editor of IEEE Transactions on Advanced Packaging. Dr. Tentzeris was a Visiting Professor with the Technical University of Munich, Germany for the summer of 2002, where he introduced a course in the area of High-Frequency Packaging. He has given more than 40 invited talks in the same area to various universities and companies in Europe, Asia and America. He is a Senior Member of IEEE, a member of URSI-Commission D, an Associate Member of EuMA and a member of the Technical Chamber of Greece.

— Prof. Manos M. Tentzeris

NSF-PRC Associate Director for RF Research School of ECE

Georgia Institute of Technology

Atlanta, GA 30332-250 U.S.A.

Printed in the United States
by Baker & Taylor Publisher Services